Robert Montgomery Smith Jackson

Flora or Vegetable Life of the Mountain

Robert Montgomery Smith Jackson

Flora or Vegetable Life of the Mountain

ISBN/EAN: 9783337268749

Printed in Europe, USA, Canada, Australia, Japan

Cover: Foto ©berggeist007 / pixelio.de

More available books at **www.hansebooks.com**

FLORA;

OR,

VEGETABLE LIFE OF THE MOUNTAIN.

> "Many such there are,
> Fair Ferns and Flowers, and chiefly that tall Fern
> So stately, of the Queen Osmunda named;
> Plant lovelier in its own retired abode
> On Grasmere's beach, than Naiad by the side
> Of Grecian brook, or Lady of the Mere
> Sole sitting by the shores of Old Romance."
>
> WORDSWORTH.

> "Who hath the virtue to express the rare
>   And curious virtues both of herbs and stones?
> Is there an *herb for that?* Oh that thy care
>   Would show a root that gives expressions!
>
> "And if an herb hath power, what have the stars?
>   A rose, besides his beauty, is a *cure.*
> Doubtless our plagues and plenty, peace and wars,
>   Are there much surer than our art is sure."
>
> "Herbs gladly *cure our flesh,* because that they
>   Find their acquaintance there."
>
> HERBERT.

THE MOUNTAIN.

BY

R. M. S. JACKSON, M.D.,

CORRESPONDING MEMBER OF THE ACADEMY OF NATURAL SCIENCES OF PHILADELPHIA;
MEMBER OF THE AMERICAN ASSOCIATION FOR THE PROMOTION OF SCIENCE;
MEMBER OF THE AMERICAN MEDICAL ASSOCIATION; MEMBER OF THE
MEDICAL SOCIETY OF PENNSYLVANIA; MEMBER OF THE CORPS OF
THE GEOLOGICAL SURVEY OF PENNSYLVANIA; CORRESPOND-
ING MEMBER OF THE ACADEMY OF SCIENCES AND
ARTS OF PITTSBURG; MEMBER OF THE LYCEUM
OF JEFFERSON COLLEGE, CANONSBURG,
ETC. ETC. ETC.

"Heaven shortens not the life of man: it is man that does it by his own crimes. Thou mayest avoid the calamities that come from heaven, but thou canst never escape those which thou drawest upon thyself by thy crimes."—CONFUCIUS.

PHILADELPHIA:
J. B. LIPPINCOTT & CO.
1860.

Entered, according to Act of Congress, in the year 1860, by

R. M. S. JACKSON, M.D.,

In the Clerk's Office of the District Court for the Eastern District of Pennsylvania

THE plant may be characterized as organic water which is polarized upon two sides, towards the earth and the air. The vegetable vesicle must, therefore, maintain two poles. While it would represent in itself the magnetic pole, it endeavors to identify itself, to obey gravity, and merge into the darkness toward the mediate point of the earth; but that it may remain a galvanic pole it becomes excited by the air, strives to become a Different and to attain the light.

Animals are entire heavenly bodies, satellites or moons, which circulate independently about the earth; all plants, on the contrary, taken together, are only equivalent to one heavenly body. An animal is an infinity of plants.

PHYSIOPHILOSOPHY.

"As sunbeams stream through liberal space,
And nothing jostle or displace,
So waved the pine-tree through my thought,
And fanned the dreams it never brought.

" Who leaves the pine-tree, leaves his *friend*,
Unnerves his strength, invites his end.

"Whether is better the gift or the donor?
Come to me,
Quoth the pine-tree,
I am the giver of honor:
He is great who can live by me.
The rough and bearded forester
Is better than the lord;
God fills the scrip and canister,
Sin piles the loaded board.

"Whoso walketh in solitude,
And inhabiteth the wood,
Choosing light, wave, rock, and bird,
Before the money-loving herd,
Into that forester shall pass,
From these companions, power and grace.
Clean shall he be, without, within,
From the old adhering sin."

EMERSON.

THE plant may be characterized as organic water which is polarized upon two sides, towards the earth and the air. The vegetable vesicle must, therefore, maintain two poles. While it would represent in itself the magnetic pole, it endeavors to identify itself, to obey gravity, and merge into the darkness toward the mediate point of the earth; but that it may remain a galvanic pole it becomes excited by the air, strives to become a Different and to attain the light.

Animals are entire heavenly bodies, satellites or moons, which circulate independently about the earth; all plants, on the contrary, taken together, are only equivalent to one heavenly body. An animal is an infinity of plants.

PHYSIOPHILOSOPHY.

"As sunbeams stream through liberal space,
And nothing jostle or displace,
So waved the pine-tree through my thought,
And fanned the dreams it never brought.

"Who leaves the pine-tree, leaves his *friend*,
Unnerves his strength, invites his end.

"Whether is better the gift or the donor?
Come to me,
Quoth the pine-tree,
I am the giver of honor:
He is great who can live by me.
The rough and bearded forester
Is better than the lord;
God fills the scrip and canister,
Sin piles the loaded board.

"Whoso walketh in solitude,
And inhabiteth the wood,
Choosing light, wave, rock, and bird,
Before the money-loving herd,
Into that forester shall pass,
From these companions, power and grace.
Clean shall he be, without, within,
From the old adhering sin."

EMERSON.

CHAPTER IV.

FLORA OF THE MOUNTAIN.

To the Naturalist the tree stands the kingly record of the triumph of the vegetable life-principle. A transcendental cell, even the imagination can scarcely conceive that from the simplest vital monad such form of loveliness and majesty could ever arrive.

The stately palm in solitary beauty, the gigantic sequoia and lofty pine spiring to the realms of the clouds, the sturdy "everlasting oak" and imperial magnolia, the banyan-fig-tree and mangrove, must acknowledge brotherhood with the humble lichen on their trunks, or the fragile parasite on their leaves, under the overshadowing unity and tyranny of the law of "organic vesicles." It is thus that the primordial formula of the tree appears to the eye of science, under the profane microscope, the ruthless knife, and that despotic *law.*

But there *are* "Trees of Jehovah and Cedars of Lebanon," (Ps. civ. 16,) signifying the spiritual man. (A. C., 776.) "Tree also signifies man; and as man *is* man by virtue of affection which is of the will, and perception which is of the understanding, therefore these also are signified by tree. There is also a correspondence between man and a tree; wherefore, in heaven there appear paradises of trees, which correspond to the affections and consequent perceptions of the angels; and in some places in hell there are also forests of trees, which bear evil fruits, correspondent with the concupiscences and consequent thoughts of those who are there." (A. R., 400.)*

"The tree is man; the effort to produce means is with man, from his will in his understanding; the stem or stalk, with its branches and leaves, are in man its means, and are called the truths of faith; the fruits, which are the ultimate effects of the effort in a tree to fructify, are in a man uses; in these his will exists. (F. 16.) Man, who is re-born, in like manner as a tree, begins from seed; wherefore, by seed in the Word, is signified the truth which is from good; also, in like manner as a tree, he produces leaves, next blossoms, and finally fruit, for he produces such things as are of intelligence, which also in the Word are signified by leaves; next such things as are of wisdom, which are signified by blossoms; and finally, such things as are of life, namely, the goods of love and charity in act, which, in the Word, are signified by fruits. Such is the representative similitude between the fruit-bearing tree and the man who is regenerated, insomuch that from a tree may be learned how the case is with regeneration, if so be, anything be previously known concerning spiritual good and truth." (A. C., 5115.)

"The tree of life signifies perception from the Lord, and the tree of *knowledge of good and evil*, perception from the world. (Ap. Ex., 739.) Trees of Eden (Ezek. xxxi.) signify scientifics, and knowledges collected from the *Word* profaned by reasonings." (A. C., 130.) "And the Tree of Knowledge signifies the pride of one's own intelligence." (D. P., 328.)

The culmination of a vast vital series, the imperial organism of that wondrous chain between death and life,— between the organic and *inorganic* worlds, "man and nothingness," profoundly significant as an emblem of expression or symbol of utterance between the *Finite* and *Infinite*, for the tree also is a type of man, and there is a "corre-

* Swedenborg.

spondence between man and the tree;" thus an indispensable element in that shining web of uses which is the universe, the mystical and scientific representations of the tree seem to be numberless.

"All life is figured as a tree. Igdrasil, the Ash-tree of Existence, has its roots deep down in the kingdoms of Hela or Death; its trunk reaches up heaven-high, spreads its branches over the whole *Universe:* it is the Tree of Existence. At the foot of it, in the Death-Kingdom, sit three nornas, Fates,—the Past, Present, and Future,—watering its roots from the sacred Well. Its boughs, with their buddings and disleafings—events, things suffered, things done, catastrophes—stretch through all lands and times. Is not every leaf of it a biography, every fibre there *an* act *or* word? Its boughs are Histories of Nations. The rustle of it is the noise of Human Existence, onward as from of old. It grows there, the breath of human passion rustling *through* it; or storm-tossed, the storm-wind howling through it like the voice of all the gods. It is Igdrasil, the Tree of Existence. It is the Past, the Present, and the Future; what was done, what is doing, what will be done; the infinite conjugation of the verb *to do.* Considering how human things circulate, each inextricably in communion with *all,*—how the word I speak to you to-day is borrowed, not from Ulfila the Mæsogoth only, but from all men since the first man began to speak,—I find no similitude so true as this of a Tree. Beautiful; altogether beautiful and great. The *Machine of the Universe;*—Alas!! do but think of that in contrast!!"*

"The incorruptible being is likened unto the tree Azwättha, whose root is ABOVE and whose branches are BELOW, and whose leaves are the vĕds. He who knoweth that, is acquainted with the vĕds. Its branches growing from the three Gŏŏn or qualities, whose lesser shoots are the objects of the organs of sense, spread forth some high and some

---

* "Heroes in History;" Thomas Carlyle.

low. The roots which are spread abroad below, in the regions of mankind, are restrained by action. Its form is not to be found here, neither its beginning, nor its end, nor its likeness. When a man hath cut down this Azwăttha, whose root is so firmly fixed, with the strong axe of *disinterest*, from that time that place is to be sought from whence there is no return for those who find it; and I make manifest that first Pŏŏrŏŏsh from whom is produced the ancient progression of all things."*

Does not the oracular tree whisper to each ear the answer to the prayer it wants to hear? To the shepherd boy in the raptures of love,—love only, as when the "Milk-white thorn that scents the evening gale," breathes out *for* him *his* "tender tale;" to the poet, dreaming, it speaks of beauty and ecstasy, a wave of that sea of glittering globules which plays forever before his soul, "a flash of light in the infinite and eternal night;" to the savant, armed with microscope, it gives an invitation, beckoning forward to explore and contemplate forever; and to the pious devotee, in the fervors of devotion, is it not a "stream of consecrated glory, which heaven ardent opens, and lets down on man in audience with the Deity"?

It is thus that the Hebrew prophet's far-reaching adumbrations attain to final organic utterance in the transcendent soul of the Swedish seer through the spiritual interpretation of the Word; and thus, also, that the Myths, of Scandinavian Scalds, find soil for their roots in Scottish heads, and "Heimskringlas" and "Heroes in History" unite in the infinite beauty and significance of the tree. So, in far-off symbolisms and correspondences, in vague and shadowy but living and suggestive thoughts, does the *tree* stretch forth its roots, trunk, branches and leaves, flowers and fruit, into that more spiritual and ethereal world the consciousness of *man*. From ancient bibles and vedas, in inspirations of Hebrew and Hindoo bards, from mysterious Druidical sha-

* Of Pŏŏrŏŏshŏttămă. B. V., page 111.

dows, and the first mutterings of poetry and song, steals out the mysterious *life-thought*, as "Tree of Knowledge," "Tree of Jehovah," "Tree of Life," the essential celestial—"and, in a supreme sense, the Lord himself,"—"Tree of Existence," wonderful Igdrasil!! and the still more mystical and divine Tree Azwattha, Symbol of the "Incorruptible Being."

Even to the first opening intelligence of barbarous and semi-barbarous tribes, there was discovered that "occult relation between man and the vegetable," as from his earliest history a reverence for trees and forests was a marked characteristic. The primitive home of the uncivilized man, they gave the first sense of protection and comfort; the first temples of the gods, groves overwhelmed him with awe, and impressed upon him veneration for the supernals.

"Who haunts the lonely coverts of the grove :
To these, and these of all mankind alone,
The gods are *sure reveal'd*, or *sure unknown.*"*

Old in story are the woods of the Druids; old are the fables of Pan, and trees sacred to the deities of the forests; and ancient are the groves of Silvanus and Dodona. The love of woods, then, comes as a revelation of the profoundest instincts of the soul; for by no accident could appear this constant fidelity, this inevitable worship.

The retreat of the SAVAGE, the home of the POET, the temple of the PRIEST, the ancient faith and primeval worship of NATURE, was a phasis of man's development stretching down to necessary and immortal affinities, rooted in inevitable placental relationships, sacred as bonds of a divine maternity, and is still *inseparable* from the *duration* of his *normal life*, as air from his lung or blood from his heart. The forest must continue to be the heaven of ecstasy for contemplation and worship, and the haven of rest for the

* Rowe.

wounded and wearied from the dusty roads and burning fields of the world; and while the sacred retreat of the devotee of religion and beauty, they invite the sick and suffering in body and soul, the lacerated and riven in spirit and heart, to wander in their life-renewing shades. Why were the temples of Æsculapius built in groves and on mountains outside of towns and cities? A profound wisdom looms forth from the institutions and rites of the ancients, and dear perpetually to the gods is the soundness of the bodies and souls of mortals; the classical dream of Æsculapius and his daughter Hygeia shining as the prophecy of the light of true science dawning and to beam forever.

Leaving the poetry, symbolism, and far-off spiritual significance of the tree, turn to the tree itself. Botanists have distributed the trees that grow on the surface of the planet into a series of belts or zones; as "certain climatal conditions are requisite for the growth of trees, there exist certain portions of the earth's surface destitute of woods, chiefly on account of cold. The tree-limit illustrates this.* At the north this limit is sometimes 71° north latitude, and in the "southern hemisphere it extends as far as the continents."† These zones are named, commencing at the north, 1st, The zone of conifers; 2d, The zone of amentaceous or catkin-bearing trees; 3d, The zone of multiform woods; and 4th, The zone of the rigid-leaved woods.‡

These belts are again designated, by others, the zone of conifers, the zone of deciduous, and the zone of evergreen woods.§

By examining this highly interesting and attractive subject, it will be discovered that with the geographic distribution of plants is connected the whole destiny and progress of the human being, and if "necessity is (*not*) the mother of the world," she *establishes eternal limitations to all things,* and is at least that dread power that fixes the fates of men.

* Schouw. † Idem. ‡ Idem. § Schleiden.

Within the Tropics or Zone of Multiform Trees a boundless exuberance of vegetative force, with endless diversity of structure, prevails. Rich and varied in color of inflorescence and foliage, the forests of the equatorial regions are masses of life and light. The closely-packed trunks of an extensive variety of eccentric and beautifully-formed trees are chained into continuities of woods by interwoven masses, or networks of vines, which knot and rope the whole together, while their bodies, branches, and leaves are alive with parasitic plants clinging to their surfaces, or trailing in pendant festoons from stem to stem. From the disposition of the leaves, and whole style and character of the foliage, a ghastly light permeates every recess of these forests, which are also filled with a corresponding multitude of animal forms, revelling in the heat and glare which constitute the horrors of the woods of the tropics. With this light and splendor, this flaunting array of fantastic figures and brilliant coloring, the forests of the temperate zone present a most entire and perfect contrast. Leaving the brilliant but noxious display of the vegetation of the torrid spaces, the change to the cool recesses of the mixed woods of the temperate climates is one of the most striking phenomena of Nature. These forests are composed principally of deciduous trees, as the oak, beech, chestnut, maple, etc., with smaller trees mixed, and sometimes with different species of the coniferæ. They have frequently a bush-underwood or heath-growth beneath the larger trees, which is composed of a number of interesting plants, but presents nothing like the labyrinth of vines and smaller shrubs that fill the interspaces of the forests of hot climates. This is explained by the deeply-shading foliage of these woods obstructing the light from their recesses, so that few plants can grow beneath them for want of heat and light, the great life-elements of the tropics. With markedly distinguished features, this zone of plants is widely separated both from the belt of "rigid-leaved," the multiform, and the woods of the exclusive coniferæ. It contains some of the most imposing and interesting forms within the tree-limit. Sometimes single species occupy extensive surfaces, almost to the exclusion of others, their groves stretching in dense and serried ranks over large spaces.

Again, a number of species grow together,—plants widely separated in botanical affinities,—as, for example, the cone-bearers and deciduous tribes, thus giving all the elements of variety and graceful combination to this order of woods. These mixed growths of trees are of surpassing beauty, some of them exhibiting a grandeur and solemnity found only in the dark recesses of the magnificent forests of the temperate latitudes. They are the great woodlands, possessing so much value as reservoirs of timber, for all purposes, and whose importance to man, in *every aspect*, it is impossible to compute.

The Alleghany Mountain, in Pennsylvania, in its botanical developments through planetary affinities, falls, in its general relationship to the world of vegetable-life, into this belt or zone of geographic distribution of plants.

The history of this life and its diversity of types, or the "Flora" of the mountain, especially in the department of trees, is one of extreme attraction.

As the direct and necessary consequence of the geography and geology, or soil and climate of a region, as already shown, the vegetable world unfolds itself by its own fixed and unalterable laws. Next to topographic distribution of surface, hill, valley, mountain, and stream, "the vegetable clothing makes the distinctive features of a country, the tree-world, or arborescent vegetation, being especially concerned in imparting *expression* and *character* to *surfaces*."*

The mixed soils of the different tracts of the Alleghany range, in Pennsylvania, and its mixed climate from elevation above the level of the sea and the medium latitudinal geographic position in the temperate belt of the planet, marks the meeting of separate vegetable classes, and gives great diver-

* Schouw.

sity of phytological life. There is, in this region, the combination of two zones of plants just described, namely, the Terebinthinate trees, (coniferæ,) or those possessing slender stems, of great height, and needle-shaped leaves which are evergreen, with the exception of a few species, and the zone of amentaceous trees, which are plants with spreading branches and diffuse spray, bearing wide, tender, and membranous leaves, which drop every year and leave the stems and branches bare through the frost months. The mingling of these two belts, which are representative worlds, and unite widely-separated chapters of the history of the planet, with laws of media, soil, and climate, distinct and peculiar, gives a special charm and interest to the forests of the Alleghany.

A notice of some of the most striking trees composing these forests may not be found uninteresting. This recitation need not be made in the strictly scientific order of the botanist, but in the natural succession in which they might be supposed to attract the attention of the traveler. A catalogue of the most commonly observed and extensively distributed plants of the mountain, including the several departments of botany, will be appended to this chapter. The object of this must be obvious, especially to the physician, to whom the great laws of "Habitats," and the dread necessities which superintend the devevelopment and perpetuation of *Life* in all its forms, reveal themselves in the character, qualities, entire nature of the proper legitimate earth-children rooted in and united by bonds of parental affinity to special localities and in special media.

The trees and woods of this range of mountains have some distinguishing features, all of which will be apparent after a special portraiture of them shall have been made.

The mountain is clothed with an extensive and beautiful variety of trees. In their distribution upon the surface, these trees seek the most congenial localities, affecting the soil and exposure made healthful and agreeable by oldest affinities and home sympathies. Rocky height or rugged ravine, alpine table-land or sloping mountain vale, have each their primal clothing of vegetation. The southeastern slopes of the mountain, also its range of summit-knobs, are generally covered by a variety of oaks, chestnut, firs, and pines, and a number of other trees, those with deciduous leaves greatly predominating. In winter the aspect of this side of the mountain is stark and bare, the monotonous gray of the forests destitute of foliage, prevailing, with occasional spots of pines, their dark-green hue visible at all seasons of the year. Other parts of the mountain, especially the ranges of depressions of the western sides, on the contrary, show great extents of evergreen forests. Its eastern slopes and summit in full summer costume present an unrivaled array of verdure in an endless ocean of leaves,—the foliage of the hardier members of the oak family, as the chestnut oaks, with white, red, and black oaks prevailing. To these may be added the chestnut, beech, and several varieties of birch and maple, with linden, poplar, cucumber, hickory, and walnut.

As the forest is composed of an aggregate of individual trees, and the exact mode of growth of the individual giving at last a general character to the forest, some notice of the manner in which the different trees of the mountain grow may assist in the truthful rendering of its woods. This special portraiture of trees, or the study of the growth and mode of development of each kind of tree, properly belongs to the artistic department of natural science, and is especially attractive to the naturalist who is not a mere cataloguemaker. Besides the exhaustless beauty of the variety of form, and the special attraction of specific styles of growth, this study of the individual tree reveals great laws of science in the necessities which superintend the unfolding of its structure. This style of growth is thus a theme of twofold attraction, interesting to botanical physiologists, and especially interesting to the artist or student of form,—for the different varieties of trees have forms and expressions as different and characteristic as the separate races of animals; indeed, each individual tree, flower, or rock, is a unit as perfect as any other unit, whether animal or man. All men are more or less alike, so are all blades of grass; all trees are more or less alike, so are the birds in their branches. But the artist who works from Nature had better forget his patterns or stereotype trees of different orders, his *model men, birds, and plants;* for in the *living universe* they are *all,* also, exceedingly *unlike each other.* Each man and grass-blade is an *individual* having all those modifications of external or secondary qualities which mark him or it from all other men and grass-blades, and makes it that *individual,* unlike all *other men* or grass-blades of the universe. So must the *real artist* paint the individual tree; and thus is the world endless in opulence of resources, and each form of each new tree is a study, and its integrity and beauty renewed forever. Hence, also, is this worshiper in "God's first temple" enveloped in a perpetually new atmosphere of light and loveliness; and thus does he drink from fresh rivers of ethereal wine, and in the deep beatitude of the artist's love of beauty feels that he could be entranced for a thousand years.

No contrast can be more striking than that which exists between the evergreen trees and the deciduous, or those which assume only a summer dress, being arrayed for occasions. Their forms are as differently suggestive as the substances which constitute their bodies. Different members of the pine family affect the shape of the pyramid, yielding to the imagination the idea of duration, by giving a base which no storm can uproot or turn over, their tapering summits, at the same time, presenting the smallest surfaces for the attacking winds. The oak and the beech, very different from the pine, fling out their arms into wide, umbrageous, overshadowing masses of limbs and twigs, which only seem to wish to grow on and cover the largest space. Thus the pine-tree sings its song and has its dance of joy in the war of the winds, and the tempest's roar is its frolic, while the branches of the oak and beech are whirled and twisted like withs in its fury, their leaves being torn to rags and scattered like dust in the tornado's path. . One class represents the hardy pioneers of a world in a process of reclamation from chaos, for

> "My garden is the cloven rock,
> And my manure the snow.
> And drifting sand-heaps feed my stock,
> In summer's scorching glow."

Thus a full-armed warrior, ready for battle at a moment's warning, stands the pine-tree; the other the representative, also the fruit of a riper time, belonging to a more progressed system, requiring richer soil and fatter provender, can only sport destructible leaves for a short time, soft and evanescent, and requires constant protection and care.

Something of the individual tree or species, its style of development or architecture; something of the fashion of tree-building on the Alleghanies, may introduce the inquirer to a clearer recognition of the laws of organic life under the despotism of physical conditions and the grave necessities of habitats.

In noticing the trees of the mountain, without reference to scientific classification or precedence, we commence with the white ash, as a representative of use and beauty. This is the FRAXINUS AMERICANA of the botanists, and is certainly a family connection of Igdrasil, the Ash-tree of Existence, but just where, in botanical, natural, or artificial systems, is not recorded.

The books quote it as "a large tree, fifty or sixty feet high." This description will not apply to the tree as it grows on the mountain. It there frequently attains to five feet in diameter, with a height of 120 feet; its close-ribbed, deeply and finely sulcated light-gray bark covering a trunk as straight as a granite shaft sometimes for eighty feet, and without a branch. At this height it separates into branches, forming a head of finely divided limbs and spray, its small, green pinnate leaves pubescent and glaucous beneath in 3–4 pairs, giving to its delicate foliage an expression strange and unsuitable for a tree of such majestic proportions. One other less stately species of this genus grows on the mountain, the FRAXINUS PUBESCENS.

MAGNOLIA ACUMINATA.—This is the mountain magnolia, or cucumber-tree. Beck describes it as a "middle-sized tree, sometimes, however, attaining the height of seventy feet;" and Darlington represents it as a "majestic and symmetrical species, sixty to eighty feet high," which would convey but a remote idea of the proportions assumed by this splendid plant on the Alleghanies. It frequently exhibits a diameter of four and a half feet, with a beautiful undivided stem of ninety feet, as straight as a plumb-line, covered by a laminated white bark, with narrow but not deep grooves, the whole tree attaining the height of 120 feet. The leaves of this tree are of exceeding beauty, dark-green and glossy above, and beneath bluish and pubescent, often twelve inches long by six inches in width. The flowers are large and handsome, but not gayly colored, and are followed by a reddish fruit, like a small cucumber, possessing a highly aromatic taste and smell. It is found in considerable abundance in the depths of the forests on the western sides and table-lands of the mountain, and grows mixed with other trees. Its style of foliage and growth gives the tree a peculiar and distinguished cast, its large dark-green leaves attracting the eye, as if some majestic stranger had wandered into the forest; so exotic and foreign in its aspect that the beholder is reminded of tropical palms and mangroves. The lumber of this tree classes in value with that of the poplar or tulip-tree. Gray suggests that, "possibly the Magnolia Fraseri (the long-leaved cucumber-tree) grows in the mountains of Pennsylvania." He also quotes the Magnolia umbrella as being found on the mountains of Pennsylvania. On the Alleghanies they have not yet been seen.

ACER.—This is the family of maples, several of which are found on the mountain.

ACER SACCHARINUM, or the sugar maple, grows here into a large tree. Both varieties (the saccharinum and nigrum) are found growing sometimes five feet in diameter and 110 feet high. Its trunk is rough and twisted, with rugged, scaly bark, when it grows in open woods, but slender, straight, and smoother when it grows in deep forests with other trees, or in dense groves of its own species. The white silvery wood of this tree is much valued as fuel, also for cabinet purposes, especially when that freak, or "fantastic trick" of the woody fibre occurs, producing what is called "bird's-eye maple." Its well-known sugar-sap gives one of the staples of the mountain.

ÀCER RUBRUM is also found here. This species is called "rock maple," and furnishes the variety of cabinet lumber called "curly maple."

The ÀCER PENNSYLVÁNICUM, striped maple, moosewood, or striped dogwood, is a small, slender tree, with beautiful foliage, and dark-green, handsomely-striped branches. It grows abundantly on the mountain, but has no value as timber, its trunk never attaining more than a few inches in diameter.

"ÀCER SPICÀTUM" is a tall shrub which grows in clumps and thickets in the gorges and ravines of the mountain. This little plant is called "mountain maple," and, although only a bush, it bears a most striking resemblance to its imperial brothers, the arborescent species.

There are several indigenous cherry-trees on the mountain. These are of the genus CÉRASUS.

CÉRASUS PENNSYLVÁNICA is a graceful little tree, quoted by the botanists at twenty to thirty feet high, but often twice that height. It bears snow-white blossoms on thin, bright-red, and purple branches and twigs, followed by a red, sour little cherry.

CÉRASUS VIRGINIÀNA is the choke-cherry. This is rather a bush, scarcely ever aspiring to the tree form, and grows along streams, bearing abundance of astringent fruit on short, close racemes.

The CÉRASUS SERÓTINA is the wild black-cherry, said to grow "thirty to sixty feet high." On the Alleghanies this plant is a superb tree, often five feet in diameter and 125 feet high. It grows in groves or mingled with other tall trees, and rivals the tallest of them in height. When it grows in this manner it exhibits the peculiar shape of the mountain trees growing in dense woods. This form of trees has been brought about by the circumstance of their original growth. The mass of foliage rises in a plane forming the tops of the trees. As the lower limbs become shaded and atrophied, they die and drop off, and at last branch-buds cease to be developed and branches to grow, the trunks or stems extending upward as naked symmetrical shafts of mathematical regularity, the terminal branches forming a leafy summit or canopy, which continues to mount higher and higher as the mass of the forest rises in the air.

In open woods and low-lands of the State, this tree grows in a widely-spreading umbrageous mass, the stem dividing into a number of branches, the whole tree scarcely attaining half the height of the same plant struggling in the depths of the mountain forests. A stem without a branch for ninety feet, and as straight as a gun-barrel, is a common form of the plant in these woods. This is the "cherry-lumber" tree, so much valued as cabinet material.

Of the allied genus PRUNUS, the mountain has one species, the "Americana." It has the ordinary characters of the tree elsewhere.

BEECH-TREE.—Of the genus FAGUS, the continent, and, consequently, the Alleghany, has but one species, and that is the FAGUS FERRUGINÉA, or American beech. The mode of growth of this tree in the mountain forests is so entirely different from the shape of the tree elsewhere that it seems to have lost its identity. This is so markedly the case that common observers have made several beeches of the botanical *one species*, as "white" and "red," "mountain" and "water" beech.* These varieties are of course produced by

---

* When the heart-wood, (*duramen*,) which is a flesh-red color, is large in proportion to the white, (*alburnum*,) or sap-wood, it is called

the circumstances of growth, as of the soil, air, moisture, and other special surroundings of the plant. As it grows on the mountain, the tree, an object of loveliness, is especially attractive, and it would seem that in the beech the spirit of grace and beauty had found its most appropriate image and symbol of perfection. It grows in extensive continuous forests, the rugged web of interwoven roots forming almost a floor for miles, while the white symmetrical stems, uninterrupted by branches to a great height, present the appearance of Grecian columns, giving an expression of art to these vast and leafy sylvan temples. In striking contrast with the hemlock forests, the beech groves appear in gay and fanciful antagonism. They grow everywhere on, but seek the flatter slopes of the mountain, and seem to affect the gentle undulating surfaces of the table-land. These forests, with their series of white columnar trunks sporting long, thin, and graceful branches, covered with delicate, green, membranous leaves, half translucent, present an array always festive and beautiful. In the early spring, when the tender tissue-paper young leaves are unfolding, and present their soft and delicate surfaces to the air, it is hard to imagine anything more ethereal and exquisite than a waving grove of this lovely tree. In autumn, when the leaves have turned yellow, they appear almost to possess a self-luminous or phosphorescent power, for, at this time, however dark the night may be, or dense the forests, the traveler sees his path illuminated by a mild, diffused light, each object integrated as by a hazy moon or snow. The effect of this mystic and peculiar light is enchanting. After being for a time in beech woods the contrast is fearful, if the pathway lies through a hemlock

"*red beech.*" This occurs when the tree is old, but with small diameter, the annual layers being *very thin*, and the limbs and foliage small in quantity and proportion. With a large amount of limbs and leaves the white wood predominates, and a tree of a given diameter may exhibit only half the number of concentric annual sheets, and be of only half the age, of a red or heart-wood tree of the same dimension of trunk.

forest. Almost perfect darkness seems at once to reign, and the journey must be groped through as in a region of absolute night. In mixed forests of these two trees the effect is always charming in the extreme, as they suggest different orders of associations and reveal different phases of the elements of life and beauty.

Of the Cone-Bearers, or Pine Family, there are not many species on the mountain. A few pitch pines (*Pinus rigidia*) and yellow pines (*Pinus mitis*) on the eastern declivities and summits, also an occasional spot on the western slopes, together with the white pine and hemlock, which are very abundant on the whole range, constitute the representatives of the evergreen, or terebinthinate order of plants.

Genus ABIES.—On the Alleghany proper there is but one species of Spruce in great abundance. There are several species of this genus on the parallel Appalachian ridges and intervening elevated valleys of Pennsylvania. Asa Gray cites this State as the locality of several species of Abies, viz., the Frasèri, Nigra, Canadènsis; and it is in the well-known botanical range of the "Balsamea" and "Alba." Some rare localities contain several of these beautiful species, with the Hackmatack. One of these localities is a delightful little "garden of the blest" among the "seven mountains" of Centre and Huntingdon counties, called the "Bear Meadows." It is a small, elevated synclinal trough, surrounded by high, sharp, white sandstone (Formation 4) mountains on all sides, with one outlet or gorge, through which flows the stream draining the valley. It is evidently the bed or rich bottom of a mountain tarn or lake, the waters of which have escaped by a rupture of the wall surrounding it. A wild, exquisite, and secluded spot, it would seem to be the fantastic Arcadia of some dreaming artist or lover of nature, hidden from the world's vulgar gaze, and consecrated to beauty. Fresh glimpses of green carpet-spots of prairie, with osier beds and clumps of stately, solemn evergreens, black, silver, and balsam firs, with pines, cedars, and laurels, open into vistas of tall, deciduous trees, artistic and surprising in their exclusiveness and grace. A dark amber-colored stream, the water stained from vegetable infusion, and exhibiting throughout the year the tint of the mountain waters during the fall of the leaf, wanders, with a thousand curves and foldings, through wastes of reeds, sedges, azaleas, alders, and andromedas, cardinal flowers, "vegetable satyrs," and the more imperial Orchis, with weird Sarracenias and goldthread Coptis. An ancient and deserted garden of rare and lovely evergreens, varied shrubbery, and beautiful flowers, this little valley seems, in its isolation and sequestered beauty, to be a fragment of Paradise left unprofaned, to remind us of the splendor of the pristine home and of glories departed.

ABIES CANADENSIS.—The Canadian fir-tree, familiarly known as the "hemlock" of the mountain, is a very abundant species. It delights in northern exposures, as if seeking to battle with the coldest winds, asking no sympathy from the more genial gales of the south. It forms large forests, thick and compact, taking a savage and exclusive possession of the surface, and destroying all other forms of vegetable life beneath them. These hemlock forests have a striking and unique appearance, unlike the forests of any other tree. Like the gloomy isles of dark, half-subterranean temples, enveloped in sepulchral gloom, the wanderer feels, as he treads their lonely and sequestered solitudes, that the darkness of night surrounds him at noonday. In sleepy silence, with hushed footsteps, he treads their labyrinths of majestic columns as if veritably in the "land of shades." In the winter they assume an extremely sombre aspect, appearing, in very cold weather, the ground being covered with snow, as if smoked or painted black. Like the forests of the "Inferno," gloomy and peculiar, the tree has a funereal hue, and chills while it invites and offers the protection of its shade. It seems exclusive, and holds its title to the surface by actual possession for hundreds of years.

The hemlock of the mountain grows sometimes to an enormous size, frequently attaining the circumference of 20 feet,

with a height of 130 feet. These vast towers of woody fibre are the records of ages of labor of the vegetable life-powers, reclaiming the carbon, earth, and water of the world from chaotic floating. They fill the observer with astonishment, their massive forms, "like pillared props of heaven," suggesting the limbs of "Atlas, whose brawny back supports the starry skies." Their scraggy and rugged trunks give more the idea of rocky shafts than trees, and, like granite needles or stone obelisks, they seem to say they will stand forever. The lumber of this tree is of great value.

PINUS STROBUS.—The white pine, like the hemlock, is scattered over the whole mountain in almost every position, rocky height, or ravine, but only prevails in extensive continuous groves along the valleys of the streams, or the cold undulating surfaces of the table-lands. It grows in dense close-set masses, which have an expression, *sui generis*, from the specific shape or style of the tree. It is the loftiest of our indigenous trees, quoted by some of the books at from 80 to 100 feet, but in primitive mountain forests its straight thin columns often attain a height of nearly 200 feet, with an exceedingly narrow diametric base. These small, tapering stems look like masts of ships or lightning-rods, their delicate hair-leaf foliage giving the appearance of green mist in their tallest boughs, the whole woods waving like a grove of colossal plumes in the wind. The sharp, tapering summits of these trees do not intercept the rays of light as occurs in the interlocked canopy of the hemlock forests, but give a green and airy lightness, diffused through their densest groves, without the oppressive sense of shade and darkness which prevail on surfaces covered by their more gloomy brother.

When the white pine grows scattered in forests of other trees, it does not shoot up in single thin stems, but frequently forks or divides into groups of stems, which spring generally from a single, massive, knotty stump, or short trunk, which rises alone from the earth. The size of this basis or pedestal of the miniature forest above is often of enormous dimension and exceedingly irregular in contour, but it is evidently most generally the result of the germination of single seeds, some of them exhibiting, however, the appearance of two or three seeds having germinated in contact. The philosophy of the growth of this particular form is, apparently, that the different species of trees forming these groves have started from the earth's surface at the same time, but somewhat scattered, and, when the first branches of the infant pine were developed, the surrounding growth prevented a lateral expansion of the limbs, each of the primitive branches afterwards becoming a separate trunk or tree, and projecting itself upward, as the pine does in other crowded forests. From the point of separation at the forks, the limbs, each a noble tree itself, spring together, frequently of one size, like an immense chandelier, and rise in the air, the whole bundle of stems being supported and nourished by one large root-base. Many of these *forked-pine* trees have quite a celebrity, and have attained the character of *individuals*, and are visited as curiosities of the mountain. The lumber of the white pine is of great value, and forms one of the chief staples of the mountain.

ULMUS.—Along the flats of some of the streams the elm often attains to a great size, sometimes dividing into regular clumps of thin trunks, which bend outward from the centre, the whole summit being flat, and the tree of the shape of an inverted bell. Three species of the genus Ulmus grow on the mountain, viz., the "*Americana*," the "*fulva*," and "*racemosa*." They seek, as elsewhere, with their characteristic instinct, the moist flats and neighborhood of streams. Many of these elms are of enormous size, and of exceedingly fantastic and eccentric forms, appearing to have, by some sylvan sorcery, been led to violate all sober and common-sense laws of tree-building, and to have grown by freaks of the vegetative forces into "monsters of such frightful mien," that, to be *remembered*, "need but to be seen."* To have introduced the photographic tran-

* An exact and perfectly-elaborated portrait of an elm of rare and grotesque form and immense proportions has been painted for the

scripts of some of these trees into ANY "Midsummer Night's Dream," would scarcely have been "to have stolen the impression of the fantasy" of Shakspeare's most ultra imaginative creations. Surely, in the presence of one of these fanciful forms, one would say, that the artist who should forget or deny that a tree can be an *individual*, in its contour and lineaments as specific and unique as a statue of Phidias or a church of Michael Angelo, had better drop his pencil, or satisfy his aspirations by transcribing the stereotype trees of his first lesson-book, and by the transference to his canvas of the pictures of fence-rails or timber-posts.

With a sense of shame the forgiveness of the wood-gods must be implored for having neglected so long one of their special admirations, a true splendor of the vegetable world, the LYRIODENDRON-TULIPIFERA, or tulip-tree, sometimes also called the *wild poplar*. The mountain sports this plant in a state of greatest perfection, its trunk attaining the largest proportions by the species anywhere achieved. A proud and lofty monarch of the American woods, it is admired as a beauty of the earth by all who have seen it. With a broad, lobed, and truncated smooth leaf giving a spe-

American Academy of Music of Philadelphia, by Russel Smith, the well-known American landscape-painter. Every limb, twig, and almost every leaf, of this remarkable tree has been fixed on a canvas forty feet square, by the wonderful power and genius of this gifted artist. The original of this picture stands on the "everglades," or what was originally the beaver-dams of one of the tributaries of Clearfield Creek, three miles northwest of the "Alleghany Health Institute." It may not generally be known, even to Pennsylvanians, that many of the finest artistic combinations in the magnificent scenography of the Academy, the grandest histrionic temple in the world, were taken from the recesses of the Alleghany Mountain forests, in their native State, by the magical pencil of Russel Smith, a *native artist*. To more intimately and thoroughly study and *work from* the beautiful models of the mountain, Mr. Smith has secured a rural cabin and piece of land near the "Alleghany Springs" and Pennsylvania Railroad summit, where, in his own words, "some of the brightest and grandest moments of my life have been passed."

cial feature to its foliage, it bears a superb flower, as rich and varied almost, in its tints and style, as the tulip of the gardens. "Erect as a sunbeam" its stem sometimes shoots into a splendid shaft, almost a hundred feet in height without a limb, and then branches into a kingly diadem, or veritable crown of leaves and flowers. This shape has been frequently described already as *peculiar* to the trees growing in deep forests. In the tulip-tree it is perhaps the most striking of all. As its enormous trunk, fluted by a deeply-grooved bark of a silver-gray color, carries an almost unaltering thickness throughout its entire height, the imagination requires no assistance to behold its mass of verdure and beauty growing from the summit of some majestic marble column "on Grecian wold." It is one of the valuable *lumber-trees* of the mountain.

TILIA AMERICANA.—This is the linden or bass-wood, "lime-tree," and "white wood," a beautiful and noble tree, attaining to over 100 feet in height. In deep groves it has also the characteristic form of mountain trees, that is, with tall, straight, branchless stem, terminating in a mass of boughs, spray, and leaves, which, together with its smooth, graceful trunk covered by white finely-ribbed bark, presents one of the most striking and beautiful denizens of the woods. The species "*heterophylla*" is also found on the mountain. The lumber of the linden sells under the name of "*whitewood*," with poplar and cucumber.

BETULA.—The Birch Family have several representatives here. These are the BETULA LENTA, "*nigra*," "excelsa," and "*papyràcea*." Some of them grow into large trees, as the "*lenta*" and "*nigra*," which are often found ninety feet high. The wood of the well-known sweet, or cherry-birch, the "*lenta*," is valuable, giving a fine-grained red lumber, and good fuel.

QUERCUS.—Several of the oak group are found here, and among them the QUERCUS ALBA, or familiar and valuable white oak. It does not grow in these localities, as in the Appalachian valleys, in continuous groves, but is found mixed in forests of other trees of the noblest proportions, which it ever assumes in any soil.

QUERCUS MONTANA affects the eastern slopes and summits, having a taste, as its common name indicates, (*rock chestnut-oak*,) for rugged and stony surfaces. Associated with the last species is the "*castanea*," and scattered in different localities over the mountain are the "*nigra*," "*tinctoria*," "*coccinea*," "*rubra*," and "*falcata*." From the size to which many of these species grow, it would seem that here must be a special home of the oaks.*

CASTANEA.—The chestnut has a special affinity for the mountain. The CASTANEA VESCA grows here to a prodigious size, living ages. It bears the familiar well-known sweet nut, and has an extremely rugged bark, covering a coarse-grained, light wood, especially prized for its *indestructibility* as a fencing material. The CASTANEA PUMILA or chinquepin, grows here also.

NYSSA MULTIFLORA.—The tupelo, black or sour gum, grows sparsely over the mountain, presenting its ordinary characters in other localities.

PLATANUS OCCIDENTALIS.—The American plane, sycamore, or buttonwood, is found on the streams at the base and on the table-lands of the Alleghanies, but not on its

---

* In connection with oaks, a word on the progressive instincts of the Pennsylvanians may be in place. It has arranged itself on the record that it required the lumber-men of Maine to come to Pennsylvania to show her mountaineers the value of their forests, the "Yankee stave-cutter" having been a pioneer in one of the most valuable lumber specialties of the mountain and the State. Transcendent Yankee!! his sharpness is past finding out; he cuts the "trees that twist with the sun," saying, that those which "twist against the sun will not hold *molasses*." Curious problem in the philosophy of *kinks;* it seems that the *refractory saccharine* principle of the South requires a special twist of a special Northern oak to hold it level, and this, too, by the *special twist* "*with the sun*," and not "*against it*." When will Pennsylvanians wake up to the special twists of Northern Fanatics and Southern Salamanders, both *with* and *against* the *sun?*

summits. Its snow-white stems, mingled with the sombre hemlock, forms one of the finest and most striking contrasts in nature. Its lumber is valued for some purposes.

The MORUS RUBRA, or "*Mulberry-tree,*" grows in perfection here.

JUGLANS.—The species cinerea (butternut) of this genus is found in great abundance along the streams in the gorges, also higher up the mountain. The "*nigra,*" or "*black walnut,*" is also found, but not in such quantity.

CARYA.—There are several species of *hickory* on the mountain. The "*alba,*" or shell-bark, grows here with its usual characters, but is not abundant. The "*sulcata*" grows along the base of the mountain, and in the little valleys of the streams. The "*tomentosa*" and "*microcarpa*" are here, but not abundant. The species "*glabra*" is very common on the Alleghany and some of its parallel ridges, constituting quite an article of commerce, the young, tough sprouts being sold for hoop-poles in immense quantities. The "*amara*" is also found here. Several of this genus are but small and insignificant trees on the mountain.

POPULUS.—Of this genus there are several species, as the "*tremuloides,*" or "aspen," the "*grandidentata,*" or long-toothed aspen, the "*cándicans,*" and "*heterophylla.*" These are graceful and attractive trees, generally with smooth stems and beautiful foliage, but do not grow in the deep forests with the large, rough, mountain trees.

The ROBINIA PSEUDACACIA, or common locust, grows in profusion on the mountain. It frequently achieves the proportions of a considerable tree, and is *valuable* as an *indestructible timber.**

SALIX.—A number of willows have made their home on the mountain, both of the tree and bush form. Several of

---

* On the question of its indestructibility, see Canal Commissioners' Reports generally of the Portage Railroad of the State, on the eternity of locust crossties, under the jurisdiction of the ship, the horses, and the plough, overshadowed by the protective wings of the American eagle—Virtue, Liberty, and Independence.

the trees are familiar and handsome, but more of the genus are plain unostentatious shrubs.

These are the principal trees which are found in the forests of the Alleghany Mountains in Pennsylvania. The poet might make a book of biography, and the artist a gallery of paintings of these splendid trees alone. An enumeration of some of the most striking trees or larger forms of the vegetable world is all that has been produced as the living mantle or robes of life and organic appendage to the mountain, viewed individually, and in the concrete, or masses of woods. What fills with amazement the explorer of these forests is the thickness or density of the growth, and enormous size of the trees. He is troubled to conceive how these huge and thickly-planted trunks, which seem to have scarcely room to stand, are nourished, or grow in the limited space allotted to each tree. Such pyramids of wood might be supposed to require some base to support them, but the trees are so crowded, that were not the surface of the earth the chained continuity of interlocked roots that it is, they *could not* stand. Where the axe commences to fell these forests, and trees are left standing alone, they soon fall to the earth for want of the support and protection of the surrounding mass. The woods are so dense as to be almost impenetrable, the under-growth frequently having disappeared entirely, the branchless and naked trunks, supporting, only on their summits, a canopy of leaf-bearing branches. In some of these forests the fallen stems of immense trees, that have died of old age, half cover the ground. Here, in deep shadows and silence, sleep the monarchs of the forest, silent and sequestered, the dark solitudes furnishing a suitable graveyard for these heroes of a thousand storms, each one reposing as he fell, for now

> " Low lies the plant to whose creation went
> Sweet influence from every element ;
> Whose living towers the *years* conspired to build,
> Whose giddy top the morning loved to gild."

They sleep, while ever-busy Nature clothes each prostrate form with a shroud of verdant mosses,—thus it is

> " Out of sleeping, awaking,
> Out of waking, asleep;
> *Life death overtaking;*
> Deep *underneath deep!*"

The general aspect of these forests, with their different changes in the procession of the seasons, must strike the most careless observer. During the winter they are stark and stern, the evergreen forests affording but a gloomy contrast, their dark-green foliage scarcely suggesting the thought of *life*, while the ceaseless moan of the cold and naked stems speaks only of death to the wolfish winds.

Occasionally, in the winter forests, a phenomenon occurs of surpassing wonder. This is the sudden transition or transmutation, frequently during the night, as if by some magical power, of the whole forest of trees into a forest of glass. The mists, rains, and air charged with moisture, invest the tree-trunks, branches, and twigs with a clothing of ice, clear as crystal, so that the woods seem invested with an unrivaled splendor. This glittering and phantasmal array must be seen to be appreciated or conceived.

The phenomenon of the hoar-frost is allied to this glass metamorphosis. This is the investment of each finest fibre of the woods with a snowy, crystalline, and sparkling velvet of frost, the air being filled with floating and brilliant spangles, detached by the slightest breath of wind.

The vernal change is most genial and striking. After the long death-sleep of the winter, as is the case in northern latitudes, the leaves and flowers, with the first sun-fires, flash out upon the air with an endless succession of tints, forms, and outlines. The shades of green of the young foliage are numerous, giving a different appearance to each newly-arrayed tree. Each plant is peculiar in the character of its new-born leaves; sometimes, as in the case of the beech, dropping from the twig a soft and delicate membrane that floats like a cobweb in the air; again, as in the chestnut, hanging sullenly as if wilted; or, again, as in the oak and maple family, obtruding their more angular leaflets, which stick out rigidly from the terminal twigs. Each tree has a form or physiognomy for its *new-born* leaf, also for the perfect organ or full-grown leaf, and these different aspects show trees as entirely unlike each other, in the different stages of unfolding, as the callow bird in the nest is unlike the full-plumed and perfect adult. Even the grave evergreens assume a new countenance in the spring from the protrusion of their annual growth of twigs which are covered with exceedingly delicate light-green leaves, giving to the tree, at this time, a gay and cheerful look. This fresh livery of the vernal forests forever inspires with joy and hopefulness; for it is the time when the world and the soul are full of promise. With electrical enchantment the spirit of the woods reaches the spirit of the man, and he expands and vibrates with the budding and unfolding leaf, "for man is one world, and hath another to attend him."

The vernal *sounds* of the woods are also striking and characteristic, appropriate and fit, as are all the harmonies of the wild. The soft, young leaf has not yet arrived at firmness enough to rustle or creak, and the boiling, simmering, far-off storm and ocean-sound is not distinctly heard at this season. A soft, muffled whisper, a wavy, stifled murmur, is all that the wind can make, the delicate, drooping leaflet having no vibratory consistency, and consequently the accumulated sound is a simple, monotonous breathing of the air through the moist, sappy lungs of the forest.

As the foliage is perfected, and the summer change comes on, the whole leaf-garment assumes an entirely different expression. The monotonous dark-green of the fully-developed summer-dress of the trees gives the wood, with its different plants, a more uniform aspect. In full array the forest is certainly richer and grander in this display of the life-powers, but it lacks the variety of the vernal tints. The color of all the leaves gradually darkens in hue as they are perfected for the execution of their work,—the nourishment and re-creation of the tree. This darkened tint is gradually increased as the leaf hardens and approaches its death-hour—the arrival of the frost. Some time before this, however, the woods present, for an interval, a sameness of feature, as if the leaves were silently at work, and had no time to give to the phases of beauty, but were hurrying up the execution of their function to pass away into the sleep of death. At this time the full-grown, hard, and stiffened leaves give to the woods the sounds or characteristic summer-voices,—that seething and singing which is the result of infinite friction and vibration of the hard, turgid, and perfectly developed foliage of *all* the trees. The roar of the woods, that great respiratory murmur, has now assumed a tone that cannot be mistaken, and the storm-winds can "howl with the voices of all the gods." The hour of dissolution arrives as the autumn approaches. At this season a change occurs, the most extraordinary of all in the life of the leaf, and gives to the forests of the mountain a richness of expression, an endlessness of variety unrivaled upon the earth. This first touch of the destroyer is, perhaps, the most extraordinary phenomenon of the whole vegetable-world, and, indeed, the most wonderful aspect which Nature reveals.

> "So fair, so calm, so softly seal'd,
> The first, last look by death reveal'd,
> Before Decay's effacing fingers
> Have swept the lines where beauty lingers."

The pageantry of the American forest in autumn has ever been the theme of the poet's song and subject of the painter's pencil. It is exhaustless, as beauty is ever that freshwater jet, that divine *halitus*, that ever-living sap of existence, circulating *up* "from the far-away centre of all things," and which each moment of time *creates* for the soul a rapture, brightly renewed forever. As this element of Nature is *intangible, ethereal*, and cannot be *appropriated*, it is consequently, to the spirit of man, unattainable, inexhaustible, divine.

It is especially the Alleghany Mountain which reveals the perfect and perpetual wonder of the American autumn,—a chapter of the beauty of the world for which the *old continents have no parallel*, and the earth's surface but *one* such spectacle. This comes of the extensive variety and mixture of deciduous trees, also of the mingling of this numerous class with the evergreen trees, in the woods of the mountain. Each tree has a regular series of colors, or *hues* and *shades of color*, through which its leaf passes, after the *death-stroke* of the frost. These are of an endless variety, and of the most extraordinary *brilliancy*. The solar spectrum is exhausted in this fantastic display of colors. A single tree sometimes stands a pillar of fire, or a glittering cloud of gold and purple, while again, the crimson blood-dye is succeeded by a tree which has taken its hue from the gaudy yellow of the nasturtion's cup, or the "dolphin's back of gold."

Thus the brilliant and diversified phenomenal has taken its most gorgeous robes from the tints of the autumnal forest. These phantom-pictures, like the other multiform phases of the woods, are transitory, and soon pass away, this whole world, vivid and flashing, being remembered like the pomp and pageantry of some splendid dream. Once seen it can never be forgotten. To the bright coloring of the groves gradually but quickly succeeds the russet hue of the *dead* and *withered leaf,* the dark-*brown*, in which it moulders away into dust. At length the death-dirge of the vanishing foliage is sung, and the monotonous gray of naked trees, relieved only by the dark-green of the pines, is the color of the woods, while the ceaseless whistle of the winter winds chills the heart with the thought of that coldness which shall know no warmth, and of that *sleep which shall know no waking.*

The forests of the Alleghany, in utility and beauty, are as exhaustless as its rocks and coal, its ocean of air, and streams of water, and present a chapter of ceaseless and perfect attractions.

THE UNDERWOOD, BUSH, OR HEATH-GROWTH OF THE MOUNTAIN.

The transition, from the regular forest-tree to the shrub or bush, is gradual. That portion of the forest which is called heath, or coppice, is composed of true woody plants— that is, plants formed of woody fibre, with perennial roots and stems, and either evergreen or deciduous leaves.

On the mountain the representatives of this department are numerous. Some of these plants have the dimensions of small trees, but never grow to what are called forest-trees, and many of them are of exceeding beauty, and some of value. Where the growth of other larger trees permits it, they form clusters highly ornamental, filling the mid-air spaces of the taller trees with an array of foliage sometimes in fine contrast with the leaves of the larger varieties. As a class, they are comely and attractive, and occupy spaces that seemed otherwise to be vacant. Many of them belong to classes of larger trees, and have been already enumerated among them, as the smaller species of Àcer, Cérasus, but never grow beyond a few inches in diameter of stem. To them, in the descending scale, succeeds the order of true shrubs. Of this class of small, woody plants there is an extensive and diversified field. A perfect catalogue in this department would be an agreeable undertaking, but such a task could not be attempted in a running schedule. A few of the prominent species are all that can be enumerated now, and, to commence with some of the smaller plants connecting the bush of the mountain with the forest-tree, take the genus OSTRYA. This is the hop-hornbeam, or iron-wood. The species VIRGINICA grows here, often achieving forty-five feet in height. The hard, compact wood of this little tree is useful, and the tree graceful in its form.

CARPINUS AMERICANA is an allied plant, smaller in dimensions, and with smoother bark, but resembling the Ostrya in foliage, inflorescence, and fruit.

CRATÆGUS.—Several of the hawthorns grow on the mountain. The shape of this hardy little scavenger family is uniform, whatever may be its locality. On the mountain it shows its usual *noli-me-tangere* roughness of thorns and scraggy branches, bright, beautiful blossoms with pleasant odors knotty, and blood-red fruit, etc. As they are not used for hedges or anything else, they seem, like many other objects, to exist for beauty and sweetness *alone*. The species here are *coccinea, tomentosa, crus-galli*, and *punctata*.

CORNUS.—Species "*Flórida*" of this genus is found on the eastern slopes of the mountain, but not on the summits or western sides in any quantity. Its flashing white flowers are occasionally seen in the ravines, where the plant grows with other trees. It exhibits its usual characters.

The "*Sericea*" and "circináta" are also found there. The "Canadénsis," or "dwarf cornel," is found on the parallel Appalachian ridges.

CÉRCIS CANADÉNSIS, or Judas-tree, is found sparingly distributed, low on the slopes of the mountain. It is a small, handsome tree, showing its usual characters.

ARALIA SPINOSA is a low, rough little tree, called sometimes the "devil's club." It grows abundantly in a variety of localities on different parts of the mountain. Its large, prickly, pinnate leaves, and rugged spiniferous stem, has so strange an expression as to attract much attention and remark.

ALNUS.—Species "*incàna*," is a small tree, often twenty-five feet high. It is found along the streams of the tablelands.

Leaving the small trees, and descending to the bushes proper, we are presented with a large number of interesting plants on the mountain. The most distinguished group of this smaller class of plants which does not exhibit the tree form, is the order ERICÁCEÆ, or Heath Family. Some of the genera of this order are deciduous plants, but of extreme beauty and splendor of flowering. Others are evergreen, and give a characteristic expression to woods of which they are the undergrowth.

Of the SUB-ORDER I. VACCINEÆ, or Whortleberry Family, the mountain has a number of genera.

GAYLUSSACIA.—Of this genus there are three species, "*resinòsa*," "*frondòsa*," and "*dumòsa*," on the Alleghany, the two first named being highly esteemed for their delightful fruit.

In many of the mountain districts the huckleberry is considered not only innocent as an article of food, but to be endowed with certain medicinal properties. Many of the bare knobs and barren heights of the mountain are covered by the whortleberry, giving varieties of fruit, which ripen in different seasons. The family is very hardy, requiring only the "drifting sand-heap" for a resting-place.

VACCINIUM, or bilberry. Species "*Macrocárpon*," or American cranberry, is found on many of the parallel mountains, and in some localities on the Alleghany itself, (from report,) but is not abundant. The *stamineum, Canadénse, Pennsylvánicum, pállidum, fuscátum*, and *corymbósom*, grow there. Some of these species are tall, graceful bushes, twelve feet high, and bear large, black and blue berries. The large, delicious, "*blue huckleberry*," is obtained from the *Pennsylvánicum*. The *Vitis-Idœa* bears a red, flesh-colored berry, but bitter and acid, without much flavor. Some of these "big huckleberries," as they are called, are found in moist places, others on dry hills and open woods.

SUB-ORDER II.—ERICINEÆ, or Proper Heath Family. Tribe, ANDROMEDEÆ. Two genera of this tribe are small, creeping plants. They are the *Gaultheria procúmbens*, creeping wintergreen or mountain-tea, and the *Epigœa repens*, ground-laurel. The genus ANDRÓMEDA contains a number of handsome bushes, and one tree, *arborea*. The species on the mountain are the *calyculáta, racemósa, Mariána, Ligustrina*, several of which are tall and comely shrubs.

Some of the Andromedas are found in moist, barren spaces or sandy tracts, and, like the whortleberries, seem to have strong affinities for desolate and unreclaimed wastes. Some of them are found in sphagnous swamps, and altogether realize, in the habits of growth of some of the species at least, the spirit of the poetry of the name they bear, "the fabled exposure of Andromeda the unhappy."

Tribe RHODOREÆ.—Several genera of this noble tribe are found on the Alleghanies.

AZALEA, False Honeysuckle.—There are here several species of this plant, generally fine flowering bushes, with a brilliant array of colors in their inflorescence, and possessing delicate odors, which fill the air of the woods with a charming perfume. They are extensively distributed over the mountain, and are frequently mingled in dense brakes or heaths of other bushes of the same natural order.

AZÀLEA ARBORÉSCÉNS is a fine bush, twelve feet high, bearing large, red, fragrant blossoms.

"VISCOSA," Clammy, or White Honeysuckle, is here. It is a beautiful shrub, ten feet high, with white rose-tinted flowers in large clusters, which are very fragrant.

"NUDIFLORA," or Purple Azàlea, is a bush five feet high, bearing a purple and showy flower. It is one of the handsomest species, and has a great many varieties.

The "CALENDULACEA," or Yellow Azàlea, also grows on the mountain. It is a tall bush, twelve feet high, bearing orange-colored blossoms, and giving brilliancy and light to the copse where it grows.

Of evergreen shrubs or bushy-plants of this order, the "KALMIA" and "RHODODENDRON" are the principal.

The "KALMIA," or American Laurel, is a well-known plant, growing on all the mountains of the Middle and Northern States. It is much esteemed for its richly-varnished evergreen leaves, and its splendid array of delicately-tinted flowers. It frequently grows in dense brakes in cool, moist forests, forming what are called "laurel thickets."

The KALMIA LATIFOLIA, or Mountain Laurel, abounds on the Alleghanies, and is the only species of the genus found here. In the dense thickets in which it grows it is frequently seen twenty feet high, with long, knotty roots and twisted stems. It grows abundantly on almost every part of the mountain, and is found in immense continuous brakes, frequently under dense masses of forest-trees, seeming not to be affected by the absence of light in such places. It bears a profusion of beautiful white and rose-colored flowers, which are much admired. The leaves and fruit of this plant are poisonous.

But by far the most beautiful individual of this order, the real pride of the mountain, is the "RHODODENDRON," or Rose-Bay tree. This splendid plant, which is generally called "big laurel," is not a laurel, but closely allied to it. It belongs, with the laurel, as we have just seen, to the family of heaths, or natural order Ericaceæ, sub-order Ericineæ, and tribe Rhodoreæ. It differs from the laurel very essentially, forming a separate genus called Rhododendron, the proper botanical name of the laurel, as has just been stated, being Kalmia. Unlike the latter, it is not poisonous, and differs in its foliage and inflorescence, being a much more imperial and distinguished plant. With the common laurel it covers considerable tracts of the mountain forests, and, like that plant, it seeks the cool, sequestered shades of the deepest wilds, preferring the banks of mountain streams and unfrequented places. A splendid savage, he lives upon the sand-soil in the roughest parts of the mountain, flourishing, like an imperial chief of his order, in unapproachable seclusion. Sometimes, with the common laurel, it forms dense groves, called "*laurel swamps*,"—very improperly, however, as they are not *water plants*, and will not grow in swamps. Together, they form *thickets*, so dense and interwoven that it is almost impossible for man or animal to pass through them, thus making a wall as impenetrable as a Mexican chaparral. They have been, from time immemorial, the terror of the huntsman, as his life was in danger if he attempted to penetrate their inextricable labyrinths.* They are also the horror of the husbandman who has the audacity to attempt to clear the surface where they grow; but especially are they the trouble of the surveyor, who, with transit and compass, axe and chain, intrudes upon them. Many a youthful engineer will remember the days of his chain-carrying and rod-fixing through these thickets, and how frequently he found himself enveloped to the chin by a net of iron thongs, which held him like the jaws of an insidious trap.

The style of growth of the Rhododendron is peculiar. The stems writhe and twist themselves together in every conceivable shape of knots and tortuosities, and wherever the branches touch the ground they strike root, and the plant grows afresh from this point; it is thus that an interlaced web of stems, almost as stiff and hard as iron, is stretched over large extents, which are as impassable as cane-brakes. The traveler who attempts to traverse these thickets finds himself continually caught by loops and dead-falls. The lover of the beauty of the woods, however, will find in these sylvan labyrinths, these evergreen seas of living plants, an attractive department of the Mountain Flora. During the inflorescence of this plant it is impossible to conceive of anything more splendid than its mass of flowers, which are borne in large showy terminal corymbs or clusters. They are of a pale-rose color, and sometimes snow-white, the greenish throat of each blossom being spotted with yellow or red. Its large, thick, coriaceous leaves frequently attain the length of a foot. During winter and in intense cold, they fold or coil up longitudinally, each leaf showing a roll not much larger than a cigar, which drops down close along the terminal twigs of the plant. When a

---

* The engineer corps who located the railroads across this mountain chain, discovered the skeletons of several men who had been lost and starved to death in these thickets of Laurel and Rhododendron.

branch in this condition is carried into a heated chamber, the leaves may *be seen expanding* and rising from the close compact bunch, and assuming the flat or patent attitude, the points stretching upward as in the summer air. Under the influence of very severe frost, with the folded condition of the leaf described, the plant exhibits the same blackened, gloomy appearance, which characterizes the evergreen trees under the same conditions.

In full foliage and inflorescence the Rhododendron stands the monarch of the American heath, and always impresses the beholder with emotions of delight, from its presenting a striking contrast with the more homely and familiar forms of the other tribes of bushes. Of this less imposing, but graceful and beautiful department of the mountain forests, constituting the true underwood of the woodsman, there are many plants which are objects of attraction.

The HYDRANGEA ARBORÉSCENS is a bush found in many parts of the mountain. This plant, like some of the evergreens described, seeks the gloom of the depths of forests, its white flowers and dark-green leaves, in shady ravines and woody solitudes, effecting a perpetual surprise.

HAMAMELIS VIRGINICA, the "Witch Hazel," a tall shrub, is here, as elsewhere, a common plant. Late in the autumn its yellow flowers may be seen among the dead and withered leaves of other plants, affording a strange and startling contrast with the surrounding forms, blooming, when their blossoms and foliage are dead. This "weird" shrub stands the noblest symbol of the true and loving heart, blooming with promise and joy in the midst of desolation and death.

CORYLUS AMERICÀNA and ROSTRÀTA are found in the mountain. They grow on its slopes and the vales at its base. This filbert group seem to have an affinity for the mountain.

DIRCA PALUSTRIS.—This plant grows abundantly along the streams in ravines and small vales. It is interesting on account of the peculiar kind of bark of the plant,

which is as tough as leather, consequently, called "leather-wood."

EUONYMUS ATROPURPUREUS, or Spindle-tree, is found here. It is a showy shrub with waxy, crimson fruit hanging by long fruit-stalks.

CEANOTHUS AMERICANUS grows on the mountain, in certain districts quite abundantly. It affords excellent browsing for the deer, and is the plant used by the soldiers of the Revolution for tea.

ROSA.—Several wild roses abound, as the Lucida blanda and Carolina, with the introduced species Rubiginosa and Micrantha.

RUBUS.—Several species of this interesting genus flourish here.

RUBUS ODORATUS, or Flowering Raspberry, grows in great abundance and in the finest proportions. It shows a profusion of splendid purple flowers from June until August. These flowers often exhibit a disk of two inches in diameter, and are of great beauty. The Strigosus and Occidentalis are found with their usual characters.

The VILLOSUS, or High Blackberry, is found in the greatest quantity. This hardy bramble flourishes wherever any kind, even the poorest and roughest soil, exists. Its fruit is produced in such abundance that it forms one of the crops of the mountain. Some varieties occur.

Species CANADÉNSIS (Dewberry) grows profusely.

RHUS.—There are several species of this genus here, as the "typhina," or stag-horn sumach, "glabra," smooth sumach, "copallina," dwarf sumach, and "aromatica," or fragrant sumach. They are handsome shrubs with graceful, delicate foliage and acid crimson fruit. The poisonous species, Venenata and Toxicodendron, are rarely found on the mountain.

TAXUS BACCATA, variety CANADÉNSIS, is the American Yew or Ground Hemlock. It is a prostrate trailing bush, found in the gorges and on shaded precipices of the mountain. It has handsome, shining evergreen leaves, and bears a berry-like fruit of a blood-red color.

VITIS.—The grape family has established itself on the mountain. The "labrusca" is found in moist places, developing a large fruit with coarse and acrid qualities. This is said to be the parent stem of the Isabella grape, a variety much improved by cultivation.

Species Æstivalis grows in great profusion.

Species Cordifolia, or Frost Grape, grows also well.

AMPELOPSIS QUINQUEFOLIA, Virginian Creeper, is found here. Its crimson foliage in autumn, clinging around stumps and trees, gives a marked feature to the woods.

CELASTRUS SCÁNDENS, or "wax-work," occurs, but is not abundant. Its yellow pods, displaying scarlet-covered seeds, are esteemed ornamental.

AMELÁNCHIER CANADÉNSIS, or Shad Bush, grows profusely on the mountain. Several of the varieties described prevail, as the "botryapium" and "oblongifolia."

SAMBUCUS CANADÉNSIS, or Common Elder, abounds.

Species Pubens, Red-berried Elder, is found in great quantities, especially on the eastern slopes of the mountain. Its bright-scarlet berries, ripening in June, are borne in great profusion, looking like bunches of blood-red coral, and mingling frequently with the array of plants in bloom at this season, the splendid Epilobium, Phlox, Lobelia, and Flowering Raspberry with broad purple petals, give variety and unrivaled splendor to these floral groups.

PYRUS CORONARIA.—The American Crab-Apple was omitted in the list of small trees. It sometimes grows to twenty-five feet in height, bearing rose-colored blossoms which possess a delightful fragrance. A variety, not so brambly and scraggy as the common crab, occurs here, with taller trunk, cleaner limbs, and much larger apple.

The Sassafras and Benzoin are also found here, the latter growing profusely.

PYRULARIA (Mich.) Oleifera, Oil Nut, is said to be found

on the "mountains of Pennsylvania, near the Alleghanies." (Mich., Gray.)

The water-courses and humid tracts show large quantities of several small species of the genus SALIX, or Willow. They fringe the banks of mountain brooks and springs, and form close, compact waving masses, or osier beds, in swampy spots. The Common Alder (*Alnus Serulata*) is also found in some places covering the banks of streams and moist places, seeking with characteristic instinct the trails of springs and fountains. With the Willow, being essentially aquatic, or lovers of water in their propensities, their presence is always the harbinger of the appearance of that element, their groves being thus the true haunts of the aquatic gods, or "Water-walkers." These plants form a beautiful and characteristic order of copse, or under-bush, their wand-like stems and peculiar foliage marking them distinctly from the other species of bush. Thus *variety*, which seems to be Nature's perpetual trick to enchant her children with forms of beauty and elements of use, here finds a stripe of newness wherewithal to demand attention and admiration. A descriptive catalogue of all the mountain copse would be an attractive chapter, but a glimpse at this beautiful department must satisfy us here.

After dwelling on the lofty and imperial dendroid forms of the vegetable world, also its royal families of smaller shrubs, with their artistic beauty and almost regal pomp of ornament and extravagance of dress, another class of plants, still less imposing, but more graceful and lovely, press upon the attention of the wandererer in the mountain woods. This is the world of flowers, so called, as if perhaps they *existed* to *flower alone*, and had no account to render of themselves, but that they were revelations of the splendor and perfection of things, and brought messages of light and gladness to the soul. Of this numerous class many are found distributed over the Alleghany. They are the fairest, frailest, and most evanescent of all vegetable forms, springing from the earth each season, germinating, flowering, and seeding, then withering and dying, having but one short summer to publish their little lives, "sparkle, and expire." These many-painted forms rise as if by magic; endless variety in unity, and unity in endless variety, is the song they sing. In this world the graces and loves seem to reign, for of the grace and beauty of posies, and the positive loves of flowers, who has not heard? Why this untold riches, why this infinite diversity of form, why this exhaustless profusion of dyes,—only for beauty, only for thought and spirit? So would sing the poet the secret of their spirit and life which "the ages have kept."

"If I knew
Only the herbs and simples of the wood,—
Rue, cinquefoil, gill, vervain, and agrimony,
Blue-vetch, and trillium, hawkweed, sassafras,
Milkweeds, and murky brakes, quaint pipes, sundew,
And rare and virtuous roots, which in these woods
Draw untold juices from the common earth,
Untold, unknown, and I could surely spell
Their fragrance, and their chemistry apply
By sweet affinities to human flesh,
Driving the foe and 'stablishing the friend,—
Oh, that were much! and I could be a part
Of the round day related to the sun
And planted world, and full executor
Of their imperfect functions."

This "sweet affinity to human flesh" is the *great fact* of their being, and is quite a sufficient excuse for their existence. The mountain is rich in its array of flowering plants, some of which are Alpine in their characters. In the order of their appearance, the first that attract attention are the vernal flowers. These, as in all high mountain localities, rush rapidly into life the moment the frost has liberated the surface from its *power*. They spring from the soil in a multitude of graceful forms. Some of them are peculiar, and belong exclusively to mountains, seeming to find their proper life-medium in the cool fresh air of elevated districts, and withering when removed from those regions. The vernal flowers are numerous in this range of heights, and to be known and enjoyed in all their sweetness, must be seen and studied in their native haunts. The snows have scarcely disappeared before the first plants put forth their leaves and delicate petals upon the cold, raw air, and are especially prized and hailed with joy by the botanist as the prophets of the coming world of life. And first in rocky nooks and dripping springs, creep out, as if fearful that the winter winds might return, the Saxifrages, Draba, Heuchera, Hepatica, and Caltha, or Marsh Marigold, in quick succession. Then follow the Wind Flowers, the Wake Robin, the Spring beauty, or Claytonia, Houstonia, and Columbine, Sanguinaria, Bellwort, Corydalis, and Erythronium; mingled with these are delicate violets of almost every hue, and of which numerous and petted family the mountain has many representatives.

The little humble earth-gem "Mitchella," soon dots the green surface with its minute snow-white twin flowers, and the lovely Epigæa, with its graceful trailing stem, and foliage like painted parchment hiding clusters of delicate flowers with faint but delightful ordor, is soon found creeping among the dead and fallen leaves as if to conceal its beauty and sweetness, and give it all to the earth upon whose bosom it clings so closely. More showy plants soon flash out their light upon the air, as flaming Phloxes, Cardinal Lobelias, the Epilobium, or Great Willow Herb, with wand of showy flowers, the proud Lily, and fanciful Orchideæ, among which are the imperial purple-fringed Platanthera, with eccentric and anomalous Cypripedium, or Lady's Slipper, the bizarre form of which remarkable flower is the perpetual joke of the woods, a shape so odd, fantastic, and unexpected, that one asks if it were not created in derision. The mountain's show of summer and autumnal flowers is equally extensive and beautiful. As the summer, or sun months, are a short season on the mountain, this world of plants seem to hurry on to the full publication of their lives, and especially to render glorious a short and brilliant career, with extravagant demonstration of ornament and show.

The scythe of the first frost finds a rich and abundant harvest. The summer plants bloom on and mingle with the autumnal flowers, which seem smiling and unconscious of their coming doom. Of the autumnal flowers the Compositæ, or compound flowers, are the most numerous. Rough and hardy, they appear at the close of the flower season, proud and defiant, as if they braved the hour of dissolution. A number of this class, after slumbering in the soil nearly through the summer, suddenly start and bloom, to be as quickly nipped and destroyed. These plants seem to defy the seasons and to have resolved that they *will*, at all hazards, *bloom*. This immense order has numerous genera* and species on the Alleghany as elsewhere. Many of them are large, showy plants, and strike the most careless observer by the brilliancy of the tints of their flowers, and jantiness of their style of growth. Some of them are the largest and most conspicuous of annual plants, and are considered rough and intrusive weeds, possessing, however, rare and real beauties, as the Helianthus, Eupatorium, Actinomeris, Heliopsis, Vernonia, Lactuca, Hieracium, etc. Others of the order are more delicate and attractive in the style of their beauty, as the Asters, or star-flowers, which present a flashing array of shining faces, radiant as jewels, and of every dimension and tint of color, from white specks, minute and sparkling as snow-flakes, to broad dark-red and azure rays, until the far-famed star-flower of the celestial empire is rivaled in its perfections.

Imperial and proud, the sun-flower (Helianthus) flaunts his colors, as if he were veritably "a son of the sun," and would shine as long as his sire. The Golden-rod, (solidago) its delicate wands studded with flowers, contributes to the "mute music," and makes gay the forest and mountain's

---

* See catalogue of genera at end of chapter "Flora."

sides in those "bright September days." The Gnaphalium, Coreopsis, and Rudbeckia, mingle their silver and gold with the pageantry which heralds the advent of autumn and waves the *first farewell* to departing summer. Winter concludes the story of the flower, and its "little life is rounded with a sleep."

Need the observation be made that the full and elaborate biography of the flowers of the mountain, with their special habits, phases of life-manifestation, and instincts, would be a labor of delight? Here, again, the real lover of nature will find that she is ever true and faithful to her accredited devotees. Coy and cruel, with a face of adamant and steel to selfishness and profane intrusion, she is approachable, gentle, and pliant, to earnestness and love. Thus it will be found, that the life and habits of one plant, read and studied with devout and careful seeking, is a key to the history of earth and air, and a pass-word to the intellectual throne of the knowledge of the realms of organic life.

SERIES II.—PLANTS WITHOUT FLOWERS.

Leaving the first great division of the vegetable kingdom, the series of Phænogamous Botany, or those having definite and clearly marked organs for the reproduction of specific forms, and descending in this chain of organisms, we arrive at another order of plants with marked and distinctive features, called the Cryptogamous, or flowerless plants, or those the *mechanism* of *whose* reproduction was formerly supposed to be concealed, or even non-extant. The first division embraces the imperial forms, the great trees of the ages, the myriads of flowers which beautify the earth; also the useful plants, the companions of man, the proper bread or human-flesh grasses, or cerealia, plants furnished with easily discoverable generative systems, and all propagated by definitely organized seeds. The other division (Cryptogamic) is a more humble series,—organizations of simpler and more homogeneous elements, revealing a less intricately complicated morphology, or, in other words, fewer of the wonder-workings of that same strange cytoblast, from "those minims of the vegetable world," single-cell plants, to the more complicated structure of tree ferns, but all propagated by spores, or simple reproductive cells.

These CELLULARES, or cellular plants, are an interesting department of the vegetable world. Here commences the mysterious circulation of "organic water," and the protean power of that magical "protoplasm," with generative fiat, starts the whirl of the brute elements through the harmonious gyrations of Life. Here the formative forces of vitality assume their simplest attitudes of nutrition and reproduction; and here the "vegetable vesicle" stands the witness of the first erotic approach of the ponderable and imponderable. This is also the realm in which the two great kingdoms, the vegetable and animal, approximate and touch circles in a series of surprising analogies, in the first simple mechanisms of life, for the cell is the result of the *ultimate* analysis of *both*.

"The starting-point of both is the same; for the embryo of the animal up to a certain grade of its development, consists, like that of a plant, of nothing else than an aggregate of cells. The lowest class of animals, the microscopical animalcula, or the invisible inhabitants of stagnant water, appear to be *identical* with the simple cellular plants, already referred to (Volvox globator.")*

"Kutzing does not admit any essential distinction between animals and vegetables.† He maintains that the same being may, at various periods of its development, assume one nature or the other. The following is his theory in a few words:—Every organic being is constituted of vegetable elements and animal elements, and, according as one or other prevails, the being becomes an animal or a vegetable;

* Goadby.
† See quotation from Robert Smith, at end of catalogue.

in the first stages of development of superior beings, and permanently in those of inferior rank, the two elements are equally balanced, and this is the case, in the author's opinion, with the Diatomeæ, which, on this account, cannot be absolutely referred either to one series or the other, but constitute the ring or circle which unites together all organic beings into one kingdom. Long controversies have sprung up between the supporters and opponents of this doctrine, who, to obtain victory, mutually accuse one another of logical errors, of sophisms, and of paradoxes."*

With the exception of the tree Ferns, (Tropical,) whose trunks sometimes attain to the height of forty feet, the cryptogamic plants are lowly structures, the feathers, hair, and microscopic down of the skin of the world.

This division of plants is constituted of three classes, viz.: the ACROGENS, the ANOPHYTES, and THALLOPHYTES. The first of these classes contains four orders, the Equisetaceæ, Filices, Lycopodiaceæ, and Hydropterides. The second two orders, the Musci and Hepaticæ; and the third four orders, the Lichenes, Fungi, Characeæ, and Algæ.

ACROGENS.

Of this class the mountain has the following representatives:—

ORDER EQUISETACEÆ, (Horse-tail Family.)

GENERA. SPECIES.

Equisetum, (Horse-tail. Scouring Rush,) 2

ORDER FILICES, (Ferns.)

Polypodium, (Polypody,) Tree Fern Family, . . 2
Allosorus, (Rock Brake,) 1
Pteris, (Brake. Bracken,) 1
Adiantum, (Maiden Hair,) 1
Cheilanthes, (Lip Fern,) 1

* Meneghini, Botanical and Physiological Memoirs, Ray Society, 1853.

GENERA.	SPECIES.
Woodwardia, (Woodwardia,)	1
Camptosorus, (Walking Leaf,)	1
Asplenium, (Spleenwort,)	4
Dicksonia, (Dickson's Fern,)	1
Woodsia, (Woodsia,)	1
Aspidium, (Shield Fern, Wood Fern,)	2
Onoclea, (Sensitive Fern,)	1
Osmunda, (Flowering Fern,)	2
Botrychium, (Moonwort,)	1

The ferns are the most showy, and generally attractive of the cryptogams. Many of them are tall feather-shaped plants, their broad spreading fronds, the ornamental and imperial plumage of the earth, producing the perpetual impression of beauty. Some of them are humble and lowly plants, but possessed of exceeding delicacy and grace. The more imposing species frequently occupy extensive spaces of the mountain heath, forming brakes, or matted continuities, which cover the surface sometimes to the entire exclusion of other small plants. These fern forests have frequently a striking and characteristic expression, from the large fronds all assuming one position, by that instinct which turns the leaves of plants to the sun. The tall plumes are marshaled in order, and stand with a gentle northern inclination, their spreading pinnæ, or leaflets, looking to the south, or facing the sun, and held in file by the strong attraction of his rays. Some of the species are shy and retiring in their habits, and are rarely seen; others are found almost as common as grass, occupying large spaces, growing in the woods and swamps, while others invest the rocks and cliffs, festooning their edges and surfaces with rare and picturesque fringes and wreaths. In most of this family of delicate and comely plants, the light and spiritual forms of the vegetable world, it would seem that *beauty*, or the transcendent element of taste was *alone* consulted; while the "homely utilities" or economical relationships, with a few exceptions, had been ignored.

The ferns seem to be attracted to mountains, and, from

the shyness of their habits, their lonely, retired haunts, in rocky nook or "bosky dell," they speak constantly of sequestered solitudes, walks sacred to the wood-gods, of the isolation and self-sufficiency of nature, and of the mountain spirit, wild and indomitable in all its forms.

The man who has no memory of fern islands mingled with his boyhood's dreams, has not yet drained all the enchanted goblets of the universe, and may have still the ecstasy of a new experience in the revelation of the delicacy and sentiment of nature in her most touching attitudes of wildness, sweetness, and seclusion.

By consulting the catalogue of genera, it will be seen that most of the prominent and interesting forms of the order Filices are represented on the Alleghanies.

ORDER LYCOPODIACEÆ, (Club-moss Family.)

GENERA.	SPECIES.
Lycopodium, (Club-moss.)	5

The species of ground pines, or club-mosses, are found extensively distributed over the mountain in shaded woods and moist places. They are among the most beautiful and striking of the cryptogamic plants.

ANOPHYTES.

ORDER MUSCI, (Mosses.)

This interesting class of cryptogamic plants is extensively distributed over the world. The greater number require a certain humidity of atmosphere, and they are more numerous in temperate latitudes than the tropics. They are lowly and minute, but graceful and beautiful, and are among the first plants which take possession of rocks and sterile soils,—appearing even on volcanic slags and lifeless earth-crusts. Many of them occupy extensive swampy tracts, (the Sphagna,) and form, by their accumulations of leaves and stems, large deposits of carbonaceous mould, (modern formula of the *coal seam*,) while others climb the highest mountains, and penetrate the coldest arctic spaces. Hardy cosmopolites! they are found wherever light and moisture can penetrate, and ornament by their graceful foliage the most forsaken nooks, crannies, and neglected places. In summer their vivid velvet-mantles and verdant cushions gleam through the forests, investing nearly every prostrate trunk or living tree, bank, rock, and bed of brook. In winter their fragile bright-green leaves may be found fresh and smiling beneath frost and ice, and their tiny fruit-bearing stems carrying fantastic caps on bursting spore-cases, are often seen penetrating the snow with a reproductive energy that defies the most intense cold. This is a floral chapter that seems perennial in its fascinations, and the bryologist is especially happy as even winter gives no interruption to his attractive labors. Mountains seem to be the special home of the moss family, as the valleys, cultivated lowlands, and prairies, do not appear to attract this little race of rock and desert-taming pioneers of the vegetable kingdom. The Pennsylvania Alleghany range is a rich and varied moss district, and has been examined, to a certain extent, by a number of cryptogamic botanists.* It seems to possess the condition of elements most favorable as a habitat of this class of plants. In its cool air, its widely extended forests with interminable shades, and quantities of fallen and decaying timber, its extents of surface covered with fragments of rock, its moist ravines and gorges with projecting cliffs, its sequestered dells and shady precipices, its swampy places and fresh running-streams,—we are presented with a medium of special adaptation to the life-affinities of the Bryaceæ. Embracing several geological formations with diversity of mineral composition, which gives origin to a variety of soils from disintegration, and to the exposure for moss-growing

* Of the number who have visited the mountain for moss-gathering purposes, are the accomplished bryologists Leo Lesquereux and Thomas P. James, two indefatigable workers in this department of science, to whom the American student of botany owes much, and, it is to be hoped will owe more, before their labors are ended.

surfaces of different kinds of rock, at the same time stretching up through a considerable calorical scale, or height of climatal *variations*, it would be natural to expect that the mountain would reveal extensive botanical affinities in this and other departments. Here, to the common observer, the variety would appear to be infinite, but the drilled eye of the naturalist soon classifies and catalogues them all.

Of the ORDER MUSCI, William S. Sullivant and his co-laborers in this department, have reported three hundred and ninety-four indigenous species, of which two hundred and fifty-five are common to Europe.

No introduced European species are recorded. Some cursory observations of the mosses of the Alleghany give the following list of genera:—

Sphagnum, (Peat-moss.)
Phascum, (Earth-moss.)
Gymnostomum.
Aphanorhegma.
Physcomitrium, (Bladder-moss.)
Hedwigia, (Beardless-moss.)
Tetraphis, (Fourtooth-moss.)
Grimmia.
Schistidium.
Racomitrium, (Shredcap-moss.)
Fissidens, (Splittooth-moss.)
Dicranum, (Fork-moss.)
Leucobryum, (Pallid-moss.)
Ceratodon.
Campylopus, (Swanneck-moss.)
Weisia.
Rhabdoweisia, (Streak-moss.)
Drummondia.
Tetraplodon, (Collar-moss.)
Trichostomum, (Fringe-moss.)
Barbula, (Beard-moss.)
Atrichum, (Smoothcap-moss.)
Dicranodontium, (Swanneck-moss.)
Trematodon.

Pogonatum, (Haircup-moss.)
Polytrichum.
Encalypta, (Extinguisher-moss.)
Orthotrichum, (Bristle-moss.)
Diphyscium.
Bartramia, (Apple-moss.)
Aulacomnion, (Furrowcap-moss)
Mnium, (Thymethread-moss.)
Bryum, (Thread-moss.)
Funaria, (Cord-moss.)
Leskea.
Thelia.
Neckera.
Cylindrothecium,(Cylinder-moss)
Leucodon, (Whitetooth-moss.)
Leptodon, (Wing-moss.)
Anomodon
Climacium, (Tree-moss.)
Homalothecium.
Hypnum, (Feather-moss.)
Fontinalis, (Fountain Moss.)
Zygodon, (Yoke-moss.)
Dichelyma.
Pylaisæa.
Platygyrium.

This imperfectly elaborated catalogue of genera embraces more than half of the described North American mosses. It will be increased, no doubt, largely by future and more critical explorations, as many of them are exceedingly minute plants, with shy habits, and requiring great patience and vigilance to discover them. These genera contain a number of species, many of which are found on the mountain, thus giving an extensive list of mosses for that locality. A number of them are of great beauty, and widely distributed. Sometimes they show matted masses resembling forests of miniature pines; again, microscopic cane-brakes, or laurel thickets, investing with their delicate tree-shaped stems the rocks and ground. Considerable spaces of the surface are grown over by some of the species, as the earth is covered by grass. Others, again, are found on trees, covering their trunks and branches, while there are those that inhabit fountains and brooks, and occupy the surfaces of rocks and fallen timber, enveloping whole prostrate trunks with mantles of variously tinted plush, or robes of delicate light-green feathers. Thus, as objects of grace and beauty, they constitute an interesting field of investigation, dressing the myriad shapes of the woods with elaborate and fanciful decorations. Being very retentive of life, and hardy, they resist extremely low temperatures, many of them fructifying, as we have seen, in the snow, and exhibiting their bright foliage when other plants are sleeping or dead. The moss thus appears a silent witness, in the slumber of winter, that life has not been extinguished in the whole world of vegetation, or animation even suspended *in all*, but that in some forms it is imperishable, blooming with the freshness of evergreen youth through all times and seasons.

Another point of interest in this class, as of other cryptogamic plants, is their world-wide or cosmopolitan range. Thus we have seen that of 394 described species of indigenous moss, 255 are common to Europe, while of the whole number of species, including Phænogamous plants, enumerated by Gray, 2668, only 676 are common to Europe.

ORDER HEPATICÆ, (Liverworts.)

This order of cryptogamic plants has many representatives on the Alleghany Mountain. They are small cellular plants, some of them resembling mosses, and presenting many points of similarity, in form and habits, with that order. They are diversified in their forms of vegetation and reproduction, the quaint and peculiar style of which is striking, even to the ordinary observer, and possesses marked attractions for the botanist. In Gray's Manual, William S. Sullivant has reported 38 genera, and 108 species. For the clever monograph, with beautifully elaborated figures of the genera of this order, by this distinguished cryptogamist, the student of American botany must feel under perpetual obligation. A heretofore comparatively closed book is now unsealed, and the student can walk with open eyes into a new and enchanting region. The Alleghany, as already remarked, is a rich Hepatic field, and will give an abundant harvest to the laborer in this department.

THALLOPHYTES.

ORDER LICHENES, (Lichens.)

The class of organic forms called THALLOPHYTES, are the simplest vegetable structures. They have no distinction into stems, roots, or leaves, as the higher cryptogamic plants exhibit, but are composed of a mass of cells accumulated in a parenchymous plane, called a *Thallus* or *Frond*. The order Lichenes is in this group. They are peculiar, both as to their nutrition and reproduction, and show a strong bearing toward that troublesome region of speculation in which commence the great questions of UNIVOCAL and EQUIVOCAL generation. On SPONTANEOUS generation, or "matter assuming organization under the influence of water and light," the following observations of Lindley may seem to savor of an unorthodox philosophy to many who are given to intellectual stampedings on the announcement of the great generalizations of science :—*

"On this subject the investigations of Meyer are exceedingly interesting. By sowing Lichens he arrived at some curious conclusions, the chief of which are that, like other imperfect plants, they *may owe* their origin either to an *elementary* or a *reproductive* generating power,—the *latter* capable of development like the plant by which they are borne: that decomposed vegetable and some *inorganic* matter, are *equally* capable of *assuming organization* under the influence of *water and light;* and that the pulverulent matter of Lichens is that which is subject to this kind of *indefinite propagation*, while the spores lying in the shields are the only part that will really multiply the species. He further says, that he has ascertained, by means of experiments from seed, that *supposed* species, and even some genera of Acharius, are all forms of the same; as, for instance, Lecanora cerina, Lecidea luteo-alba, and others, of the common Parmelia parietina."†

Of the character, habits of distribution, general nature of the Lichens, Lindley proceeds to observe :—"Pulverulent Lichens are the first plants that clothe the bare rocks of newly-formed islands in the midst of the ocean; foliaceous Lichens follow these, and then Mosses and Liverworts. (D'Urville, Ann. Sc. 6, 54.) They are found upon trees, rocks, stones, bricks, pales, and similar places; and the *same species* seem to be found in many different parts of the world: thus the Lichens of North America differ *little* from *those* of Europe. They are not met with on *decaying mat-*

---

* The mountain being a page of the venerable tome, perhaps a whole leaf of "that elder Scripture writ by God's own hand, Nature," would not desire to appear, except as witness or attorney for plaintiff in issues against those profane burglars, pick-locks, and spies, in the private workshop of the Almighty, called *men* of *science*, (wicked rogues of *nescience!!*) and the municipal corps, or regularly organized simon-pure orthodox police of Heaven.

† Vegetable Kingdom, John Lindley, Lichenales.

ter where they give way to fungi; but they often occupy the surface of living plants, especially their bark. In the tropics they lay hold of evergreen leaves. Their chosen climate is one that is temperate and moist; aspects to the north or west are also their favorite resort, for they shun the rays of the noon-day sun. No place seems to be a more constant haunt than the surface of sandstone rocks, and buildings in cool and moist countries. They are met with, in one place or other, from the equator to the pole, and from the sea-shore to the limits of eternal snow. The finest species are found near the equator; the most imperfect, such as the crustaceous genera, which can hardly be distinguished from the rocks they grow upon, are chiefly observed on mountain-tops, and near the pole. The Idiothalami are most abundant in tropical America."

The Lichen appears, then, the pioneer of that splendid world of forms which seems, from its entire dissimilarity of structure, to ignore its affinity or alliance by any conceivable nexus with it, and as the first blundering effort of inorganic matter to enter the higher sphere of life. The mountain's rocks and forests present an extensive field of research in this department of botany, from the same causes which give exuberance of growth to the other orders of cryptogamic plants. From the lower varieties of pulverulent and crustaceous lichens covering stones, fences, and walls, with white, gray, or yellow scurf, to the more complex structures of fronds, they are found picturing the surfaces of all fixed objects with every conceivable shape of spots and markings, clinging to the bark of trees, investing their branches with fantastic scales and gelatinous skins, or floating festoons of hair; destitute of roots, but eroding at last the hardest vitreous slags. Drawing their nourishment from the air, they adhere to the naked surfaces of everything, clothing the rock and tree with an endless variety of dress and ornament.

In the class of uses, the order abounds in valuable elements, nutritive, medicinal, and chemical coloring-principles.

Something of the extent of the alliance as distributed over the earth, and enumerated by Lindley, may be inferred from the fact that the species, long since described, amounted to 2500.

The ORDER FUNGI, of the class Thallophytes, is one of the most curious and interesting of the whole vegetable series. Viewed with reference to their complex structure, their strange and eccentric habits; their peculiar economic function in the organic world; their chemical composition and special relations,—they constitute a wonderful order of plants.

"A full account of the diversified modifications of structure that Fungi display, and of the remarkable points of their economy, would require a volume."* Something of this vastness of range in numbers and affinities may be imagined from the general enumeration of the "Vegetable Kingdom,"† (Alliance 11 Fungales,) amounting to 598 genera and 4000 species. The forests of the Alleghany Mountain, with their extensive variety of decomposing vegetable matter, give a large catalogue of mushrooms.

The last, but not least wonderful of the THALLOGENS, is the ORDER ALGÆ. It is a vast family, swarming in myriads through seas, rivers, brooks, and pools, and growing sometimes on wet and humid earth. The Algals are of all dimensions, forms, and colors, from microscopic points floating in water as motes swim in the sunbeam, to trailing, leathery masses, hundreds of feet in length, and from transparent mucus scums, to brilliantly tinted sheets dyed with all the colors of the rainbow. This order and its subdivisions also occupy a singular and questionable position in the scale of organisms. Of this ambiguity Lindley observes:—"It is here that the *transition* from animals to plants, whatever its true *nature may* be, occurs; for it is incontestable, as the varying statements of original observers testify, that no man can certainly say whether many of the organic bodies

---

* Asa Gray. † John Lindley.

placed here belong to the one kingdom of nature or the other. Whatever errors of observation may have occurred, these very errors, to say nothing of the true ones, show the extreme difficulty, not to say the impossibility, of pointing out the exact frontier of either kingdom."* Whereupon the Rev. M. J. Berkeley, startled "by these astounding statements," remarks:—"The same species may assume a vast variety of forms according to varying circumstances, and it is highly instructive to observe these changes; but that the *same spore* should, under different circumstances, be capable of *producing* beings of almost *entirely different nature*, each capable of producing its species, is a matter which ought not to be admitted generally without the strictest proof." In the Zoogeny of Oken it is written, (paragraph 1775,) "Every organic originates from a mucus-point. If this mucus-point occur in the darkness, it thus becomes a terrestrial organism, a plant; if it enter into the light, which is only possible in the water and in air, it thus becomes a solar organism, independent of the planet, self-moving around itself like the sun, an *animal*." "The animal is a whole solar-system, the *plant* only a *planet*. The animal is, therefore, a whole universe, the plant only its half; the former is microcosm, the latter micro-planet." (*Idem*, paragraph 1780.) So sparkle the philosophers on the origin of things, particularly of plants and animals, and all this from the contemplation of the wonderful life-manifestations of the Algals. The streams, pools, springs, and moist spots of the mountain, abound in numerous fresh-water genera and species of this widely-distributed order of plants.

Thus endeth the story of the plant. In stately and majestic repose the mountain folds about itself this many-tissued, many-tinted garment of living fibres, each microscopic alga, each imperial tree, quickened by that worker of perpetual miracles, *life*. For what ends exist this immea-

* See observations of Meneghini and Goadby, p. 250.

surable array of attractive objects? First as a vast expanse of living, normal, and beautiful forms, it shall address the senses of the physical man, and by healing sympathies and recuperative vitalizing forces, invite him to a larger, more genial, and healthful world of sensuous emotion. Secondly, it shall, by a purer, more subtle, ethereal and Divine force, penetrate the depths of his spiritual nature, and by sentiment and thought, intelligence and love, magnify and ennoble his soul.

The forest, the heath, the flower, the fern, the moss, the lichen, form thus for man a recipe of health, a concert of harmony, a lesson of wisdom, a transport of beauty.

CATALOGUE OF
FLOWERING PLANTS.

A FULL or descriptive catalogue of *all* the *species* of flowering plants would occupy more space than can be allotted to this department of the natural history of the mountain. An enumeration of the most prominent genera, or common families, with a number of the most *prominent species*, will indicate something of the predominating influences grouped under the name of habitat of the region, as shown by the plant. With the plants already named in the text, we are presented with the following:—*

SERIES I.—FLOWERING OR PHÆNOGAMOUS PLANTS.

CLASS I.—EXOGENOUS OR DICOTYLEDONOUS PLANTS.

SUB-CLASS I.—ANGIOSPÉRMÆ.

1. ORDER RANUNCULÀCEÆ, (Crow-foot Family.)

GENERA.	SPECIES.
Atràgene, (Atragene)	1
Clématis, (Virgin's-Bower)	1
Anemòne, (Wind-Flower)	3
Hepática, (Liver-Leaf)	2
Thalictrum, (Meadow-Rue)	2
Ranúnculus, (Crow-foot Buttercup)	4
Càltha, (Marsh Marigold)	1
Aquilègia, (Columbine)	1
Zanthoriza, (Yellow-root)	1

* This enumeration is the order pursued by Professor A. Gray in his admirable and invaluable "Manual of Botany." Nearly all the plants of this catalogue are common in the interior, middle, and western part of the State of Pennsylvania, as well as on the Alleghany range of mountains.

GENERA.	SPECIES.
Actæa, (Baneberry, Cohosh)	1
Cimicífuga, (Bugbane)	2

2. ORDER MAGNOLIACEÆ, (Magnolia Family.)
(Already enumerated.)

4. ORDER MENISPERMACEÆ, (Moonseed Family.)
Menispérmum, (Moonseed)	1

5. ORDER BERBERIDACEÆ, (Barberry Family.)
Caulophyllum, (Blue Cohosh)	1
Podophyllum, (May-apple, Mandrake)	1

8 ORDER NYMPHÆACEÆ.
Nuphar	1

10. ORDER PAPAVERACEÆ, (Poppy Family.)
Papaver, (Poppy,) from Europe	1
Chelidonium, (Celandine,) from Europe	1
Sanguinaria, (Blood-root)	1

11. ORDER FUMARIACEÆ, (Fumitory Family.)
Adlùmia, (Climbing Fumitory)	1
Dicentra, (Dutchman's Breeches)	3
Corydalis, (Corydalis)	1

12. ORDER CRUCIFERÆ, (Mustard Family.)
Nastúrtium, (Watercress)	3
Dentaria, (Pepper-root, Toothwort)	2
Cardámine, (Bitter-cress)	2
Árabis, (Rock-cress)	3
Barbarèa, (Winter-cress)	1
Erysimum, (Treacle-mustard)	1
Sisymbrium, (Hedge-mustard)	1
Sinapis, (Mustard)	2
Dràba, (Whitlow-grass)	2
Lepidium, (Pepper-grass)	1
Capsélla, (Shepherd's Purse)	1
Raphanus, (Radish)	1

15. ORDER VIOLACEÆ, (Violet Family.)
Sòlea, (Green Violet)	1
Viola, (Violet Heartsease)	10

16. ORDER CISTACEÆ, (Rock Rose Family.)
Lechea, (Pinweed)	1

FLORA OF THE MOUNTAIN. 265

19. ORDER HYPERICÀCE.E, (St. John's-wort Family.)
GENERA. SPECIES.
Hypéricum, (St. John's-wort).. 4
Elodèa... 1

21. ORDER CARYOPHYLLÀCE.E, (Pink Family.)
Dianthus, (Pink) introduced...
Saponària, (Soapwort) from Europe.. 1
Silène, (Catchfly Campion)... 3
Agrostémma, (Corn-cockle) from Europe................................... 1
Stellària, (Chickweed, Starwort).. 1
Cerástium, (Mouse-ear Chickweed)... 2
Spergula, (Spurrey) from Europe.. 1

22. ORDER PORTULACÀCE.E, (Purslane Family.)
Portulàca, (Purslane) from Europe... 1
Claytònia, (Spring-Beauty)... 2

23. ORDER MALVÀCE.E, (Mallow Family.)
Althæa, introduced from Europe..
Málva, (Mallow) from Europe.. 2

24. ORDER TILIÀCE.E, (Linden Family.)
(Enumerated.)

27. ORDER OXALIDÀCE.E, (Wood-sorrel Family.)
Oxalis, (Wood-sorrel)... 3

28. ORDER GERANIÀCE.E, (Geranium Family.)
Geranium, (Crane's Bill).. 1

29. ORDER BALSAMINÀCE.E, (Balsam Family.)
Impàtiens, (Jewel-weed)... 2

32. ORDER ANACARDIÀCE.E, (Cashew Family.)
Rhus, (enumerated)...

33. ORDER VITACE.E, (Vine Family.)
Vitis, (enumerated)..
Ampelópsis, (enumerated)...

34. ORDER RHAMNÀCE.E, (Buckthorn Family.)
Ceanòthus, (enumerated)..

35. ORER CELASTRÀCE.E, (Staff-tree Family.)
Celastrus, (enumerated)...
Euonymus,   "   ....

23

36. Order Sapindàceæ, (Soapberry Family.)
GENERA. SPECIES.

Acer, (enumerated)...

37. Order Polygàlaceæ, (Milk-wort Family.)
Polygala, (Milkwort).. 4

38. Order Leguminosæ, (Pulse Family.)
Lupìnus, (Lupine)... 1
Trifòlium, (Clover).. 3
Robinia, (enumerated)...
Tephròsia, (Hoary Pea)... 1
Hedysarum, (Hedysarum)..................................... 1
Desmodium, (Tick Trefoil).....................................
Lespedèza, (Bush-Clover)......................................
Stylosanthes, (Pencil-Flower)................................
Vicia, (Vetch)...
Lathyrus, (Vetchling)...
Phaseolus, Kidney Bean).......................................
Ápios, (Ground-nut)... 1
Baptísia, (False Indigo).. 1
Cercis, (enumerated)..
Cássia, (Senna).. 1

39. Order Rosàceæ, (Rose Family.)*
Prunus, (enumerated)...
Cerasus, (enumerated)..
Spiræa, (Meadow-Sweet)....................................... 3
Gillènia, (Indian Physic)....................................... 2
Agrimònia, (Agrimony)... 1
Gèum, (Avens)...
Potentílla, (Cinque-foil)....................................... 3
Fragària, (Strawberry)... 1
Dalibárda, "   1
Rùbus, (enumerated)...
Ròsa, "
Cratægus, "
Pyrus, (Pear Apple)... 1
Amelánchier, (enumerated).....................................

42. Order Lythràceæ, (Loosestrife Family.)
Cùphea, (Cuphea).. 1

43. Order Onagraceæ, (Evening Primrose Family.)
Epilòbium... 2

* See end of catalogue.

GENERA.	SPECIES.
Œnothèra, (Evening Primrose)	3
Gaùra, (Gaura)	1
Ludwigia, (False Loosestrife)	
Circœa, (Enchanter's Nightshade)	2

46. Order Grossulàceæ, (Currant Family.)

Ribes, (Gooseberry)	4

48. Order Cucurbitàceæ, (Gourd Family.)

All the cultivated members of this family flourish here, except a few of the delicate melons, which never ripen, although they grow well.

49. Order Grassulàceæ, (Orpine Family.)

Sèdum, (Stone Crop)
Pènthorum, (Ditch Stone-Crop)

50. Order Saxifràgeæ, (Saxifrage Family.)

Saxífraga, (Saxifrage)	
Heùchera, (Alum-root)	2
Mitélla, (Bishop's Cap)	1
Tiarélla, (False Mitre-wort)	1
Hydrángea, (Hydrangea) enumerated	1

51. Order Hamamelàceæ, (Witch-Hazel Family.)

Hamamèlis, (Witch-Hazel,) enumerated

52. Order Umbelliferæ, (Parsley Family.)

Hydrocótyle, (Marsh Pennywort)
Sanícula, (Sanicle)
Daùcus, (from Europe)
Angélica, (Angelica)
Zizia
Cicuta

The cultivated species of this order grow well on the mountain, as the parsley, celery, dill, fennel, and coriander.

53. Order Araliaceæ, (Ginseng Family.)

Aralia	6

53. Order Cornaceæ, (Dogwood Family.)

Cornus, (enumerated)
Nyssa,   "

Division II.—MONOPETALOUS EXOGENS.

55. Order Caprifoliàceæ, (Honeysuckle Family.)

GENERA.	SPECIES.
Lonicera, (Woodbine)	
Triosteum, (Horse-gentian.)	1
Sambucus, (Elder,) enumerated	2
Vibúrnum, (Arrow-wood)	4

56. Order Rubiaceæ, (Madder Family.)

Gàlium, (Bed Straw)	5
Cephalánthus, (Button-bush)	1
Mitchélla, (Partridge-berry)	1
Oldenlandia, (Bluets)	2

58. Order Dipsàceæ, (Teasel Family.)

Dipsacus, (introduced)	1

59. Order Compositæ, (Composite Family.)

Vernonia, (Ironweed)	1
Liatris, (Blazing Star)	2
Eupatorium, (Thoroughwort)	5
Aster, (Starwort)	12
Erigeron, (Fleabane)	4
Solidago, (Golden-rod)	13
Inula, (Elecampane) introduced	1
Ambrosia, (Ragweed)	2
Xanthium, (Clotbur)	1
Heliopsis, (Ox-eye)	1
Rudbeckia, (Cone-flower)	3
Helianthus, (Sunflower)	7
Actinómeris, (Actinomeris)	1
Coreópsis, (Trickseed)	1
Bidens, (Bur-marigold)	4
Helènium, (False Sunflower)	1
Maruta, (Mayweed,) introduced	1
Achillèa, (Yarrow)	1
Leucánthemum, (Ox-eye Daisy)	1
Tanacètum, (Tansy,) introduced	1
Gnaphalium, (Cudweed)	3
Antennaria, (Everlasting)	1
Erechthites, (Fireweed)	1
Cacalia, (Indian Plantain)	1

GENERA.	SPECIES
Senècio, (Groundsel)	1
Centaurea, (Star-thistle) from Europe	2
Cirsium, (Common Thistle)	3
Lappa, (Burdock,) from Europe	1
Krigia, (Dwarf Dandelion)	1
Hieracium, (Hawkweed)	4
Nabalus, (Rattlesnake-root)	2
Taráxacum, (Common Dandelion,) from Europe	1
Lactuca, (Lettuce)	1

60. Order Lobeliaceæ, (Lobelia Family.)

Lobelia, (Lobelia)	4

61. Order Campanulaceæ, (Campanula Family.)

Campanula, (Bell-flower)	2

62. Order Ericaceæ, (Heath Family.)

Gaylussacia, (enumerated)	
Vaccinium, "	
Epigæa, (Trailing Arbutus)	1
Gaultheria, (Wintergreen, or Mountain Tea)	1
Andromeda, (enumerated)	
Kalmia, (American Laurel,) enumerated	
Menziesia	1
Azalea, (enumerated)	
Rhododendron, (Rose Bay) enumerated	
Pyrola, (False Wintergreen)	3
Chimaphila, (Pipsissewa)	2
Monotropa, (Indian Pipe)	2

64. Order Aquifoliaceæ, (Holly Family.)

Prinos, (Black-alder)	2

68. Order Plantaginaceæ, (Plantain Family.)

Plantago, (Ribgrass)	3

70. Order Primulaceæ, (Primrose Family.)

Lysimàchia, (Loosestrife)	3

71. Order Lentibulaceæ, (Bladderwort Family.)

Utricularia	

73. Order Orobanchaceæ, (Broom-rape Family.)

Epiphegus, (Cancer-root)	1
Conopholis, (Squaw-root)	1
Aphyllon	

74. Order Scrophulariaceæ, (Fig-wort Family.)

GENERA.	SPECIES.
Varbascum, (Mullein)	2
Linaria, (Toad-flax)	1
Scrophularia, (Fig-wort)	1
Chelòne, (Snake-head)	1
Pentstemon, (Beard-tongue)	1
Mímulus, (Monkey-flower)	1
Gratiola, (Hedge-hyssop)	1
Veronica, (Speedwell)	5
Gerárdia, (Gerardia)	3
Castillèia, (Painted-cup)	1
Pediculàris, (Louse-wort)	1
Melampyrum, (Cow-wheat)	1

76. Order Verbenaceæ, (Vervain Family.)

Verbena, (Vervain)	2
Phryma, (Lopseed)	1

77. Order Labiatæ, (Mint Family.)

Teùcrium, (Wood-sage)	1
Isanthus, (False Pennyroyal)	1
Mentha, (Mint)	2
Lycopus, (Water Horehound)	1
Pycnánthemum, (Mountain Mint)	3
Thymus, (Thyme,) from Europe	
Hedeòma, (Mock Pennyroyal)	1
Collinsonia, (Horse-balm)	1
Monarda, (Horse-mint)	3
Nepeta, (Cat-mint,) from Europe	1
Brunella, (Self-heal)	1
Scutellaria, (Skull-cap)	4
Marrubium, (Horehound) from Europe	1
Stachys, (Hedge-nettle)	2
Leonurus, (Motherwort) from Europe	1

78. Order Borraginaceæ, (Borage Family.)

Echium	1
Mertensia, (Lungwort)	1
Myosotis, (Scorpion-grass)	2
Cynoglossum, (Hound's-tongue)	2

79. Order Hydrophyllaceæ, (Water-leaf Family.)

Hydrophyllum, (Water-leaf)	2

80. ORDER POLEMONIACEÆ, (Polemonium Family.)
GENERA. SPECIES.
Phlox, (Phlox) 4

81. ORDER CONVOLVULACEÆ, (Convolvulus Family.)
Ipomœa, (Man-of-the-Earth) 1
Cúscuta, (Dodder) 1

82. ORDER SOLANACEÆ, (Nightshade Family.)
Solanum, (Nightshade,) from Europe 3
Physalis, (Ground Cherry) 1
Datùra, (Thorn Apple) introduced 1

83. ORDER GENTIANACEÆ, (Gentian Family.)
Sabbátia,, (American Centaury) 1
Gentiàna, (Gentian) 2

84. ORDER APOCYNACEÆ, (Dogbane Family.)
Apocynum, (Indian Hemp) 2

85. ORDER ASCLEPIADACEÆ, (Milkweed Family.)
Asclepias, (Silk-weed) 4

86. ORDER OLEACEÆ, (Olive Family.)
Fraxinus, (enumerated) ....

87. ORDER ARISTOLOCHIACEÆ, (Birthwort Family.)
Asarum, (Wild Ginger) 1
Aristolòchia, (Birthwort) 1

80. ORDER PHYTOLACCACEÆ, Pokeweed Family.)
Phytolácca, (Pokeweed) 1

90. ORDER CHENOPODIACEÆ, (Goosefoot Family.)
Chenopodium, (from Europe) 3

91. ORDER AMARANTACEÆ, (Amaranth Family.)
Amarantus, (Amaranth,) introduced 3

92. ORDER POLYGONACEÆ, (Buckwheat Family.)
Polygonum, (Knotweed) 7
Fagopyrum, (Buckwheat,) from Europe 1
Rumex, (Sorrel) 3

93. ORDER LAURACEÆ, (Laurel Family.)
Sassafras, (Sassafras) 1
Benzoin, (Wild Allspice) 1

94. ORDER THYMELEACEÆ, (Mezereum Family.)
GENERA.   SPECIES.

Dirca, (Leather-wood) enumerated ....

98. ORDER SAURURACEÆ, (Lizard's-tail Family.)
Saururus, (Lizard's-tail).. 1

100. ORDER CALLITRICHACEÆ, (Water Star-wort Family.)
Callitriche,.. 1

102. ORDER EUPHORBIACEÆ, (Spurge Family.)
Euphorbia, (Spurge).. 2
Acalypha,... 1

104. ORDER URTICACEÆ, (Nettle Family.)
Ulmus, (Elm) enumerated..
Morus, (Mulberry) enumerated..
Urtica, (Nettle).. 1
(2 introduced.)
Pilea, (Clear-weed)... 1
Cannabis, (Hemp,) from Europe.. 1
Humulus, (Hop).. 1

105. ORDER PLATANACEÆ, (Plane-tree Family.)
Platanus, (Button-wood,) enumerated ....

106. ORDER JUGLANDACEÆ, (Walnut Family.)
Juglans, (Walnut,) enumerated ....
Carya, (Hickory,)   "   ....

107. ORDER CUPULIFERÆ, (Oak Family.)
(Enumerated.)

108. ORDER MYRICACEÆ, (Sweet-Gale Family.)
Comptonia, (Sweet Fern)..............,.. 1

109. ORDER BETULACEÆ, (Birch Family.)
(Enumerated.)
Alnus, (enumerated)...

110. ORDER SALICACEÆ, (Willow Family.)
Salix, (enumerated)...
Populus, (Poplar, Aspen,) enumerated....................................

Sub-class II.—GYMNOSPERMÆ.

111. Order Coniferæ, (Pine Family.)

GENERA.	SPECIES.
Pinus, (Pine,) enumerated	
Abies, (Spruce, Fir,) enumerated	
Cupréssus, (White Cedar)	?
Juníperus, (Juniper)	?
Taxus, (Yew,) enumerated	

Class II.—ENDOGENOUS, or MONOCOTYLÉ-DONOUS PLANTS.

112. Order Araceæ, (Arum Family.)
Arisæma, (Indian Turnip) 2
Symplocárpus, (Skunk Cabbage) 1
Orontium, (Golden-club) 1

113. Order Typhaceæ, (Cat-tail Family.)
Typha, (Cat-tail) 2
Sparganium, (Bur-reed) 2

114. Order Lemnaceæ, (Duckweed Family.)
Lemna, (Duck's-meat) 1

115. Order Naiadáceæ, (Pondweed Family.)
Potamogèton, (Pondweed) 5

116. Order Alismaceæ, (Water-plantain Family.)
Alisma, (Water-plantain) 1
Sagittaria, (Arrow-head) 2

119. Order Orchidaceæ, (Orchis Family.)
Orchis, (Orchis) 1
Platanthera, (False Orchis) 7
Goodyèra, (Rattlesnake-plantain) 2
Spiranthes, (Lady's Tresses) 1
Listera, ... 1
Arethusa, (Arethusa) 1
Micróstylis 1
Corallorhiza, (Coral-root) 1
Cypripedium, (Lady's Slipper) 3

120. Order Amaryllidaceæ, (Amaryllis Family.)
GENERA. SPECIES.

Hypoxys, (Star-grass).. 1

121. Order Hæmodoraceæ, (Bloodwort Family.)
Aletris, (Colic-root).. 1

123. Order Iridaceæ, (Iris Family.)
Iris, (Flower-de-Luce)... 1
Sisyrinchium, (Blue-eyed Grass)........................... 1

125. Order Smilaceæ, (Smilax Family.)
Smilax, (Greenbrier) 4
Trillium, (Wake Robin).. 3
Medeola, (Indian Cucumber-root)......................... 1

126. Order Liliaceæ, (Lily Family.)
Asparagus, (from Europe) 1
Polygonatum, (Solomon's Seal)............................. 2
Smilacina, (False Solomon's Seal)......................... 3
Clintonia, (Clintonia)... 2
Allium, (Onion, Garlic).. 2
Lilium, (Lily).. 2
Erythronium, (Dog's-tooth Violet)......................... 1

127. Order Melanthaceæ, (Colchicum Family.)
Uvularia, (Bellwort) 2
Streptopus, (Twisted-stalk).................................... 1
Melanthium.. 1
Veratrum, (False Hellebore).................................. 1
Helonias, (Helonias).. 1

128. Order Juncaceæ, (Rush Family.)
Juncus, (Bog Rush).. 7

129. Order Pontederiaceæ, (Pickerel-weed Family.)
Schollera, (Water Star-grass)................................. 1

131. Order Xyridaceæ, (Yellow-eyed Grass Family.)
Xyris, (Yellow-eyed Grass).................................... ?

133. Order Cyperaceæ, (Sedge Family.)
Cyperus, (Galingale).. ?
Scirpus, (Bulrush).. ?
Eriophorum, (Cotton-grass) .... ?
Rhynchóspora..

GENERA.	SPECIES.
Scleria..	?
Carex, (Sedges)..	30

This extensive genus of obscure and intricately related plants is largely represented on the mountain. In moist spots and along spring streams, pursuing the general habits of the genus, they are found in fringes and tufts, scattered almost ubiquitously over humid and other spaces. Of the 132 species contributed by John Carey to Gray's Manual, casual observations have brought into notice some thirty species.

134. Order Gramineae, (Grass Family.)

Alopecùrus, (Foxtail Grass)..	2
Phlèum, (Cat's-tail Grass, Timothy) from Europe..........	1
Sporobólus, (Drop-seed Grass)....................................	1
Agrostis, (Bent-Grass)..	1
Muhlenbérgia...	?
Calamagróstis, (Reed Bent-Grass)................................	2 ?
Stipa, (Feather-Grass)...	?
Tricúspis, (Tall-red-top)..	
Kœlèria, (Kœleria)...	
Eatònia..	
Glycèria, (Manna-Grass)...	2 ?
Pòa, (Meadow-Grass, Spear-Grass)..............................	6
Bromus, (Brome-Grass,) from Europe..........................	3
Triticum, (Wheat)..	
Hordeum, (Barley)...	
Elymus, (Wild Rye)...	1
Aira, (Hair-Grass)..	2
Danthònia, (Wild Oats)..	
Avèna, (Oat)..	
Hólcus, (Meadow Soft-grass) from Europe...................	1
Phalaris, (Canary-Grass)...	
Mílium, (Millet-Grass)..	
Panicum, (Panic-Grass)..	7
Sórghum, (Broom Corn)...	

(SEE ANTE, p. 250.)

"Given, the head of Socrates, the wisest philosopher of Greece, and a Protococcus pluvialis, a microscopic single-cell plant, is there no '*essential* distinction,' and to which does the word *incomprehensible* most justly apply? Of the creation and destiny (genesis, exodus) of a cell, or a limitless congeries of cells, (organic bodies,) of the how and why of their getting *into* special shapes or living forms from the sleep of inorganic matter, and staying there to circulate for a time within the 'ring' of *natural affinities*, then dropping *out* of that circulation into another apparently temporary sleep, called death,—or of the creation and destiny (genesis, exodus) of a man or numberless congeries of men, (Humanity,) of the how and why of *their* assuming particular styles of existence and circulating for a time within the grasp of *supernatural affinities* (supersensuous, quondam spiritual,—immaterial forces,—will, intellection, sensation, and affection, entities, real as iron or stone, but not on the chemist's table, or naturalist's catalogue,) and also, getting *out* of that material and spiritual circulation, into an apparent sleep, called, likewise, death, what has the microscopic atom, the proud mote, the wise monad, man, *the Philosopher*, to say?

"Place the dry skulls of Plato and Shakspeare beside the ruptured and effete cells of the Protococcus pluvialis and Volvox globator, and say which are the most inconceivable existences, which are the everlasting wonder of wonders. Does not the cell stand the most imposing mystery, the most incomprehensible miracle? The two problems, vast towers!! loom up from the Infinite, their summits and bases both hidden in darkness and unapproachable solitude. The broad gulf between them can only be passed upon the wings of a purer and nobler philosophy, and the deep abyss can only be fathomed by the plumb-line of a profounder and more earnest Faith."—ROBERT SMITH, *Philosophical and Religious Meditations*, vol. vii. p. 472.

FRUIT-TREES AND ESCULENT VEGETABLES.

FRUIT-TREES.

THAT the Alleghany could supply itself with fine fruit of almost every kind there is not the slightest doubt. The indifference of the mountain counties to this department of earth cultivation, as well as many other "cultures," is to be much regretted by all the friends of progress of that region. This indifference or carelessness is not confined, however, to the mountain districts of the State. The following observations of the venerable Dr. Darlington, the justly celebrated botanist of Chester County, are, it would seem, as applicable to his district as to the one here alluded to. Looking, as the inhabitants of the wilderness counties do, to the East for evidences of civilization and light, it was to be hoped that the cultivated county of Chester had passed the "*thoughtlessness*" at least, not to speak of the rudeness and barbarism deplored by the Doctor in one part of his observations on this subject. He says: "Indeed, it is melancholy to reflect how thoughtless and negligent mankind generally are with respect to providing fruit for themselves. There are few persons who do not own or occupy sufficient ground to admit of three or four choice fruit-trees and a grapevine; such, for instance, as an apricot, a peach, a May-duke cherry, a Catharine pear, and a Catawba grape; yet the great majority seem never to think of planting such trees, while they are ready enough to invade the premises, and revel on the fruits of some more provident neighbor! It is due to the *minor morals* of the community that such disreputable negligence and such marauding practices should cease to be tolerated."—*Flora Cestrica*, p. 72.

PYRUS COMMUNIS, common Pyrus, or Pear-tree.—This tree is a native of Europe. There are many varieties of this delightful fruit, which should be cultivated wherever it will grow. The mountain counties have not given the care they should to the cultivation of this tree. The seedling plant grows well on the Alleghany, and the improved varieties would of course flourish equally well. Some fine pears have been produced on the range, and it is to be hoped the subject will receive more attention.

PYRUS MALUS, Apple Pyrus, common apple-tree.—This species is also a native of Europe. Pomologists have produced and described almost innumerable varieties of this wholesome fruit. It will grow every place in Pennsylvania, both mountain-tops and valleys; but little attention has been given to this interesting department on the Alleghany. Some fine apples have been produced, and every variety and quality of that fruit can be grown there, after a time of acclimation of buds and shoots.

CYDONIA VULGARIS, Quince-tree.—This well-known tree is a native of Southern Europe. It grows well on the Alleghany.

PERSICA VULGARIS, common Peach-tree.—This member of the almond family is a native of Persia. It does not find on the Alleghany Mountain a very genial clime. Persia and the Alleghany are widely-sundered habitats, but as that mountain has a vital connection with the whole globe it must necessarily unite with Persia on some issue of fate and nature. The peach, it seems, is this happy bond, not to mention other equally interesting radicles of association!!! The juices of the fruit, as grown on the mountain, are not exactly Persian, or even Jersey-an in their deliciousness of flavor, nevertheless, it produces a peach of respectable dimension, and decidedly agreeable character. It requires constant watching and renewing by planting, as the frost frequently kills it entirely to the ground.

ARMENIACA VULGARIS, Apricot.—This delicious fruit is a native of Armenia. Very little attention is given to its cultivation in the mountain region of Pennsylvania, and on the Alleghany none.

PRUNUS DOMESTICA, common plum, Gage or Damascene.—The cultivated plums are natives of Europe.* Several of the varieties might be cultivated here with success, if attention were given to them. Those that have been tried grow well.

CERASUS, or Cherry genus.—Professor de Candolle distributes the commonly cultivated cherry into four species; Dr. Darlington and others into two. These are the Prunus (cerasus) avium, English, or heart cherry, (sweet;) and the Prunus cerasus (vulgaris,) sour red cherry, or Morello cherry. The heart cherry grows well on the Alleghany, and with a *special luxuriance* in the red shales of the eastern base and slope of the mountain. The Morello cherry also grows finely, the whole cherry family seeming to have the most friendly relations to the mountain.†

RIBES.—The current family are produced in quantities on the mountain. These are the Ribes Uva-crispa, or gooseberry, (Europe,) the Ribes rubrum, or red currant, (Europe,)‡ and Ribes nigrum, or black currant, (also Europe.) Like the native species of Ribes, the introduced species seem to flourish as if at home.

---

* Prunus domestica, L., the cultivated plum, is now deemed by the best botanists to have sprung from the sloe.—*Gray's Manual*, p. 113.

† See wild cherry, or Serasus Serotina, now Prunus Serotina, p. 220.

‡ Gray recites a "rubrum" which is found in New Hampshire as identical.

ESCULENT VEGETABLES.

Of the introduced esculent, garden, or kitchen vegetables, the mountain produces nearly all the ordinarily cultivated species and varieties. The season for growing pot-herbs, or edible plants, is short here, and also late, as the frosts of spring and fall come close together. They almost all, however, grow profusely with any care, and many varieties assume proportions which the same plant rarely attains in the lowlands. Between the valleys of the eastern and western parts of the State and the mountain heights, knobs, and table-lands, there is a difference in the time of growth and perfection of garden vegetables (this difference applying more or less to the whole vegetable world) of from two to four weeks. The results of forcing plants, as achieved in the east and west by hot-beds, hot-houses, and protected sites, is not considered in this general statement. By the use of artificial appliances, hot-beds, hot-houses, and the selection of sheltered situations in the mountain vegetables could be brought very much earlier to perfection, and grown there with the finest qualities and proportions. This subject will receive more attention, in certain parts of the mountain, soon, and extensive experiments will be made.

At the present time the farmers of that district have only small patches of a few yards in extent for kitchen-gardens, and cultivate only such plants as will grow without much care. The amount of vegetables produced in many of these little gardens is quite extraordinary, and shows that the mountain's climate and soil, with any industry, are very favorable and friendly to the class of edible plants. One point of advantage possessed by this region is, that when the staple products of the garden have passed their season, and are withered and dried in the valleys and lowlands east and west, the mountain has them green and fresh, and in the highest perfection.

The following vegetables grow well on the mountain:—

BRASSICA OLERACEA, Cabbage.—This is a native of Europe, and thrives here with several of its varieties or sub-species. These are the "acephala," or tree-cabbage, (leaves not forming heads,) the "bullata," or savoy cabbage, with finely crisped leaves, and the "capitata," or York cabbage, with *dense head*. The variety Caulorapa, (Kohl-Rabi,) bulked-stalked cabbage, grows finely, also variety "cauliflora."

The BRASSICA RAPA, sub-species "depressa," or common turnip, grows well also.

Raphanus sativus, Garden Radish.—This plant, a native of China, is hardy, and grows almost every place. There are several varieties or sub-species, as "radicula," "rotunda," "turnip radish," oblonga, common radish, also varieties of the "niger."

Hibiscus esculentus, Okra.—This plant will grow here, but has not been cultivated much. It is a native of India.

Pisum sativum, Garden Pea, and its varieties, are produced in abundance.

Phaseolus vulgaris, String Bean, common pole-bean, and Lunata or Lima Bean, grow well, but the latter will scarcely ripen on account of the shortness of the season.

The Apium Graveolens, Celery, Petroselinum sativum, Parsley, Carum and Fœniculum, Caraway, and Fennel, as already remarked in the catalogue, grow well. The Daucus carota, Garden Carrot, variety Sativa, and Pastinaca sativa, Garden Parsnip, also Umbelliferous plants, flourish equally well.

Cucumis sativus, Common Cucumber.—This plant is a native of Asia. It grows well, but the Cucumis melo, Musk Melon, will not ripen on the mountain.

Cucurbita pepo, Pumpkin; varieties do well, also the Cucurbita melopepo and Verrucosa.

Tragopogon, Oyster Plant, grows well.

Lactuca sativa, Common Lettuce, Salad, a native of India, and Helianthus tuberosus, Jerusalem Artichoke, a Brazilian plant, also flourish.

The Beta vulgaris, Common Garden Beet, has several varieties, all of which, including the Mangel-wurzel, cultivated for cattle, grow well.

Spinacia oleraceæ, Spinach, and Asparagus officinalis, Asparagus, (from Europe,) grow well.

Allium.—Several onions are easily produced, as Allium cepa, Garden Onion, Porium, Garden Leek, Sativum, Garlick, and Scœnoprasum, or Chives.

The Lycopersicum esculentum, or Tomato, grows well, but the seasons are too short to produce or ripen the fruit without a hothouse to develop the plants largely before planting out.

The Solanum melongenum, or Egg Plant, might be cultivated if the same care were taken.

The Solanum tuberosum, or Common Potato, is particularly adapted to the soil of the mountain.

Rheum rhaponticum, or Pie Rhubarb. This plant, a native of Scythia, grows luxuriantly.

www.ingramcontent.com/pod-product-compliance
Lightning Source LLC
Chambersburg PA
CBHW020330090426
42735CB00009B/1473